Human Footprints: Fossilised Locomotion?

Matthew R. Bennett • Sarita A. Morse

Human Footprints: Fossilised Locomotion?

Springer

Matthew R. Bennett
Faculty of Science and Technology
Bournemouth University, Talbot Campus
Poole, UK

Sarita A. Morse
Department of Musculoskeletal Biology
University of Liverpool, Institute
of Aging and Chronic Disease
Liverpool, UK

ISBN 978-3-319-08571-5 ISBN 978-3-319-08572-2 (eBook)
DOI 10.1007/978-3-319-08572-2
Springer Cham Heidelberg New York Dordrecht London

Library of Congress Control Number: 2014946067

Printed on acid-free paper

Springer is part of Springer Science+Business Media (www.springer.com)

For
Fredrick George Lawley (1916–1991)
Alvina Morse (1909–2008)

Preface

There is something evocative in a human track. It helps one form an instant connection with the track-maker and their journey, blending past with present. Their preservation is in itself a rare occurrence in the geological record and they contain information not only about human presence, but about the track-makers themselves, as well as the way in which they moved across the landscape providing evidence of fossilised locomotion. There is a personal connection here as well. I come from a line of geographers on my paternal side and I first worked in Scotland and the Arctic on questions of glacial geology. But my maternal Granddad – a wonderful man, who sadly is no longer of this world – was a chiropodist, or as they like to be called these days a podiatrist. So for me the study of human tracks represents a convergence in my own ancestry, one that also reflects the interdisciplinary convergence of geology with the subjects of archaeology, anthropology and podiatry needed for their study.

This is a book about human tracks, not only their occurrence around the world, but also what can be learned from them, and it aims to equip the reader with the tools to enable their study whether it be for the sheer pleasure of enquiry, in the pursuit of scientific questions in such fields as geoarchaeology and palaeoanthropology, or in the pursuit of criminals as forensic scientists. The book has been written by me as first author, but with an essential and invaluable contribution from Sarita A. Morse who while at the University of Liverpool acquired, processed and analysed much of the data described here.

Bournemouth, UK
April 2014

Matthew R. Bennett

Acknowledgements

Much of this research has been undertaken with the financial support of the Natural Environment Research Council in an award (NE/H004246/1) made to Robin Crompton (University of Liverpool) and Matthew Bennett (Bournemouth University). The support and friendship of Robin and his team Russ Savage, Karl Bates, Juliet McClymont, Emma Webster, Mary Blanchard, Colleen Goh, David Collins and Todd Pataky are all gratefully acknowledged as is access to the software Pedobarographic Statistical Parametric Mapping (pSPM). Marcin Budka (Bournemouth University) developed the Foot Processor software suite which has been used to process much of the data presented here and has contributed to the research programme over several years. Copies of this freeware can be downloaded along with some of the raw data [http://footprints.bournemouth.ac.uk/]. Sally Reynolds provided a timely and thought provoking review of the book in its final stages and we would like to thank her for her insight and advice.

Many people have provided us with access to specimens and field sites, as well as company and hard work in the field, including: Francis Thackeray, Cynthia Liutkus-Pierce, Kevin Cole, David Roberts, Gordon Roberts, Sally Reynolds, Harry Manley, David Huddart, Silvia Gonzalez, Melanie Crisfield, Dominic Strafford, Matteo Belvedere, Lisa Santello, Peter Falkingham, Gustavo Politis, Cristina Bayon, Teresa Manera and Ricardo Melchor. We would like to acknowledge the support of the Koobi Fora Field School (2007–2009) and the contribution to our research of Jack Harris, Kay Behrensmeyer, Emma Mbua, Purity Kiura and Jack McCoy as well as the numerous students of the Field School. Thanks are due to the National Museums of Kenya for allowing access to Laetoli casts in 2008, to the Iziko South African Museum in Cape Town for providing access to the Langebaan tracks and to the East London Museum (South Africa) and in particular its curator Kevin Cole, an inspiring and visionary individual, for access to the Nahoon tracks. We would like to thank Fanie Du Preez of Kuiseb Delta Adventures and Chris Lourens of the Free Air Guest House (Walvis Bay) for access and logistical support with respect to the Namibian footprints. Charlene Steele processed much of the raw field data from Namibia. The wonderful Daniel Marty offered us a range of footprint and dinosaur photographs and Cynthia Liutkus-Pierce provided information on the Engare Sero tracks in Tanzania. Former Bournemouth students Paulley, Butters, Strugnell, Perkins and Wollf are thanked for the data on track morphology and walking speed cited in Sect. 6.4. SAM would like to acknowledge her husband, family and friends for their continuous support, with special thanks to Robin Crompton for his advice and supervision. MRB would like to acknowledge the ongoing support of Bournemouth University and the contribution made by Marie Dunning, Rebecca Dolling and Kathryn Hill, as well as the support and patience of his family, whose footprints feature at various points throughout the book; the challenge now is for them to find their tracks!

Contents

Chapter 1
Fossilised Locomotion

Abstract In this first chapter we provide a broad overview of human trace fossils (ichnology) and outline the contents of and rationale for this book. The potential for human tracks to tell us about how our ancestors may have walked is discussed as is the contribution that human tracks can make in other areas of archaeology and forensic science. Key definitions are introduced, as is a simple model of human track formation.

1.1 Human Tracks

Watch couples dance or children play and you will see the foot in action; an amazing machine. Just 26 bones sheathed in skin and sinew, with muscles that can propel you forward, backwards, up and down, allowing you to twist, turn, balance and control your speed with precision. Yet despite over a hundred years of research (Morton 1935) our understanding of the human foot remains rudimentary and knowledge of how our ancient ancestors walked a subject of conjecture and debate.

Within the geological record human and animal tracks occur infrequently; freak occurrences of sedimentary preservation, with each one holding a rare glimpse of locomotive behaviour (Fig. 1.1). Currently the oldest and most famous hominin tracks are those at Laetoli in Tanzania made some 3.66 Ma ago, preserved in volcanic ash and probably made by *Australopithecus afarensis* (Agnew and Demas 1998; Deino 2011; Leakey and Harris 1987; Leakey and Hay 1979; White and Suwa 1987). In 2009 details of a track site close to the village of Ileret in northern Kenya were published as the second oldest hominin footprint site, dating to 1.5 Ma ago (Bennett et al. 2009). These footprints are believed to have been made by *Homo erectus* (Dingwall et al. 2013), one of the first species of hominin capable of long-distance walking and running. Comparison of the Ileret and Laetoli tracks has the potential therefore to explore the transition in locomotive style between the genera of *Australopithecus* and *Homo* (Raichlen et al. 2010; Crompton et al. 2012). The development of bipedalism was a critical stage in human evolution, as was the later transition from early habitual bipeds such as *Australopithecus afarensis* made famous by the skeleton 'Lucy' to endurance walkers and runners which characterise more modern humans such as *Homo erectus* and ourselves *Homo sapiens* (Bramble and Lieberman 2004). The ability of our ancestors to walk efficiently will have influenced their interaction with the

M.R. Bennett, S.A. Morse, *Human Footprints: Fossilised Locomotion?*,
DOI 10.1007/978-3-319-08572-2_1

Fig. 1.1 Modern human track made by a habitually unshod individual in fine-grained sand/silt in a dry river bed in northern Kenya. Note the: track cross-cuts ripples with heavy mineral concentrations in the troughs and the compression of these minerals in the floor of the track; rim structure formed by the up-fold of the surface laminated sands; and desiccations cracks formed after the track was made

landscape: the way they foraged and hunted for food, gathered raw materials to use as tools and their ability to migrate across the globe.

Fossil foot bones of early hominins are rarely found in association with the skeletons of known hominin species and the fossil record is fragmentary. Small bones of the foot scatter easily once released from the soft-tissue that surrounds them and consequently they are poorly preserved in the geological record prior to the advent of burial practices. But in truth fossil foot bones alone rarely give an unambiguous indication of the way our early ancestors walked, since the bones of the foot act through a series of complicated soft tissues which are not preserved. Human tracks provide an alternative source of evidence about our ancestor's feet, formed as they walked across soft-ground leaving a record of 'fossilised locomotion.' The critical question is how do tracks record the forces applied to the ground by a track-maker and what can these forces tell us about the way in which they walked? As the foot meets the ground it interacts with the substrate to leave a track which involves the convergence of biomechanics and geology.

There is also an ever growing number of human track sites discovered around the world from more recent times made by *Homo sapiens* found in such diverse settings as coastal mudflats, caves and imprinted in layers of volcanic ash (Allen 1997; Avanzini et al. 2008; Lockley et al. 2008). These sites are not only of archaeological importance in themselves, since they provide information on human presence and allow inferences about the track-makers to be made such as their stature, but they also provide reference material with which to help decipher the record of fossilised locomotion preserved within more ancient tracks. While some of these tracks are preserved in lithified, or partially lithified, volcanic ash such as the tracks on Jeju Island (South Korea), or those at Acahualinca in Nicaragua (Kim et al. 2009; Schmincke et al. 2009, 2010), most are preserved in unlithified, fine-grained silt and fine sand and in some notable cases prints are exposed by coastal erosion and then destroyed (e.g., Aldhouse-Green et al. 1993; Roberts et al. 1996). The conservation of these soft-sediment tracksites, especially when dealing with sites of palaeoanthropological significance like those at Ileret is challenging (Bennett et al. 2013). Human tracks are not only of relevance to archaeology and palaeoanthropology however since footwear evidence can in some cases be vital to criminal investigations, the proverbial 'footprint in the flower bed' (Robbins 1985; Bodziak 2000). Here geoarchaeology converges with modern forensic science with both parties having the opportunity to learn from one another.

In light of the above the aims of this volume are therefore varied and we identify four main goals: (1) to draw together in one place, a diverse literature for those interested in human tracks whether they be geologists, archaeologists, palaeoanthropologists or forensic scientists; (2) to provide a review of modern methods of data collection and analysis; (3) to explore the role and influence of substrate on track formation and preservation; and (4) to clearly state what can and cannot be inferred from human tracks. The structure of the book follows these four broad aims, but first we need to clarify some key issues of nomenclature and orientate ourselves with respect to the human foot. We recognise that those reading the book are likely to have different academic backgrounds and have therefore included a glossary located at the end of the book to aid the reader navigate any specialist terms with which they are not familiar.

1.2 Key Concepts and Definitions

Fossil footprints whether made by humans or other animals are examples of trace fossils and the technical term for a trace fossil is an ichnofossil. The study of trace fossils is therefore the study of ichnology derived from the Greek 'ikhnos' meaning track or trace. Current convention mainly derived from the study of dinosaur traces is to refer to individual footprints as tracks and a linked sequence of tracks (i.e. footsteps) as a trackway, while the track-maker is the individual who left the tracks (Table 1.1). A single, spatially-restricted track-bearing horizon is referred to as an ichnoassemblage, which becomes an ichnocoenosis if it is recurrent and an ichnofacies when it can be linked to specific sediments and environments (Hunt and Lucas 2007). There is a complex and formal taxonomic methodology for defining ichnofossils particularly where the linkage to an extant track-maker is not clear (Donovan 1994). While the formal use of ichnotaxa has been adopted recently by a

Table 1.1 Commonly used terms with respect to tracks following Marty et al. (2009)

Term	Definition
Track	A single footprint or partial impression made by the foot of an animal
True track	A track whose lower surface was in contact with the plantar surface of the track-maker's foot
Under track	A track that is formed by the compression of sediment below the track-maker's foot. When exhumed an under track may be visible but its surface will not have been directly in contact with the track-maker's foot, if for example the original contact surface has eroded. Thulborn (2012) use the term 'transmitted relief' to describe an under track which describes the situation well, but has not been widely adopted
Elite track	A well-preserved true track (Lockley and Hunt 1995; Lockley and Meyer 2000)
Trackway	A series of tracks made by the same animal (Leonardi 1987; Thulborn 1990; Marty et al. 2009)
Track-maker	The animal that made the track
Tracked surface	The surface or palaeosurface on which the track-maker walked/moved (Fornós et al. 2002)
Overall track	If the track walls – sides of a print – are not vertical then the outer track dimension (overall track) will be larger than the dimension of the track-maker's foot or the track bottom (true track; Brown 1999)
Internal overtrack	Forms by covering of the track bottom (true track) without covering the entire overall track. Often associated with the trapping of sediment within microbial mats formed in the wet print interiors (Marty et al. 2009)
Natural track cast	A mould of a track formed by infilling sediment forming a negative replica (Lockley 1991)
Overprinting	Caused by the track-maker or another animal overprinting an original track
Displacement rim	A marginal rim of a track formed by the displacement of sediment, sometimes referred to as a 'push-up' structure or a bourrelet (Allen 1997; Manning 2004)
Track ejecta	Material ejected by the removal of the track-maker's foot from a track; may be thrown forward by the track-maker's toes (Allen 1997)

few authors (Kim et al. 2008; Meldrum et al. 2011) it is not a methodology that has been widely applied to human tracksites and is not an approach that is favoured here.

At this point we need to recognise that there are different types of track and we identify three basic types:

1. Two-dimensional tracks which record the outline and surface texture of a foot; for example if one was to walk barefoot in a tray of paint one would leave a series of two-dimensional tracks until the paint adhering to the foot was removed. These types of tracks are common at some types of crime scene where a suspect or victim may leave a trail of bloody tracks for example.
2. Three-dimensional tracks which record the outline and the depth of an impression made by a foot walking on a deformable substrate. The simplest example is to think of the tracks one might make at the beach. These are the tracks which are discussed for the most part in this volume.

3. Pressure-tracks which record the outline and the contact pressure through time as a foot makes contact with the ground. There are various types of plantar force plates and pressure sensitive walkways and treadmills that record the contact pressure in various ways (e.g., peak, average, cumulative) and across different areas of the foot through time as it first strikes, makes contact with and then finally pushing off the ground. This type of information is used extensively in biomechanical and clinical studies and plantar pressure should correlate in some way with the depth of a track which in theory represents a time integrated strain response to the applied pressure.

In navigating a human track we refer to areas that reflect the portion of the foot that made it using common biological directional terms. Therefore the heel is called the proximal portion and the forefoot is the distal portion. The outside edge is referred to as the as the lateral side and conversely the inside edge is the medial side (Fig. 1.2a). The plantar surface is the bottom (sole) of the foot and the upper (superior) surface is the dorsal surface and to be consistent with this the base of a track is therefore referred to here as the plantar surface. The sides above (superior to) the plantar surface are called the track walls (Table 1.1). We describe the big toe as the first toe, also commonly referred to as the hallux. We use the word adduction to describe the situation where the first toe is in line with the longitudinal axis of the foot and abduction to describe the situation where the first toe is displaced medially. The medial longitudinal arch refers to the inside arch of the foot (i.e.. parallel to the sagittal plane) and its perpendicular as the transverse arch (i.e. running from the lateral to medial side), which follows the coronal plane. We use the term 'ball' to refer to the area proximal of the toe pads beneath the metatarsal heads and distal of the midfoot defined by the areas occupied by the medial longitudinal arch if present.

Movements of the foot in making a track are referred to by a range of terms, including: (1) dorsiflexion, the movement of the foot upwards by flexing the toes; (2) plantarflexion, the movement of the foot vertically downwards by extending the toes; (3) supination as a tendency for someone to walk on the outside/lateral edge of their foot; (4) pronation as the tendency for someone to walk on the inside/medial edge of the foot; (5) eversion as a tendency for the sole of the foot to move away from the medial/sagittal plane; and (6) inversion as a tendency for the sole of the foot to move toward the medial/sagittal plane. A wide variety of definitions and procedures are used in the literature to define the basic linear dimensions of the foot and these are reviewed in Sect. 2.6

1.3 Models of Track Formation

Figure 1.2b summarises some of the key variables which need to be considered in the formation of a human track. There is an application of a force termed plantar pressure, via the foot as it makes contact with the ground which leads to the

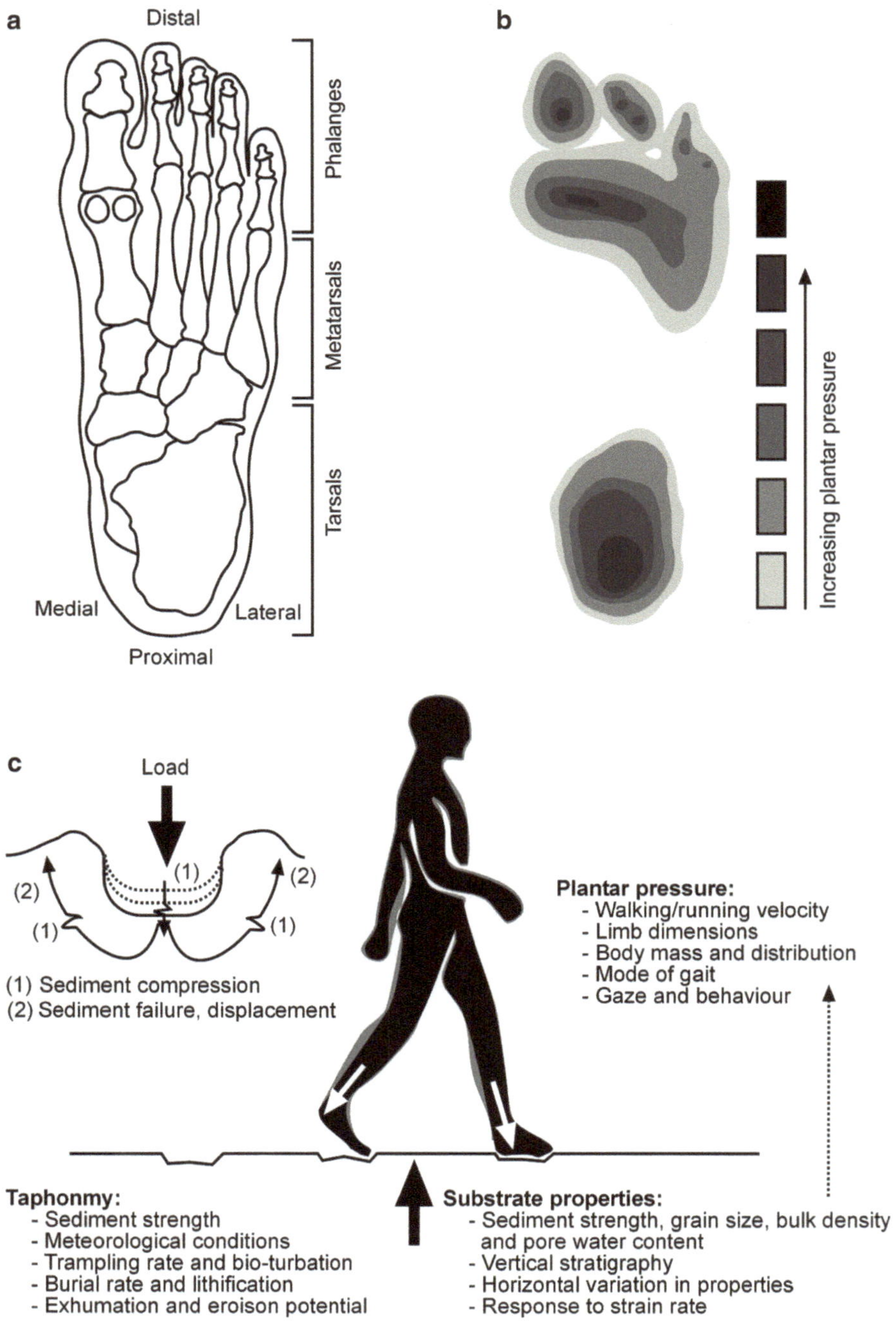

Fig. 1.2 Conceptual model of human footprint formation. (**a**) Sketch of the main bone structures in the foot, plantar view modified from Robbins (1985). (**b**) Stereotypical plantar pressure distribution associated with normal human walking. (**c**) Model of some of the variables involved in the formation of a human track. The *inset* shows the two main ways of strain accommodation in track formation, compression and deformation/displacement

compression, deformation and/or excavation of a track within the substrate assuming that the sedimentary properties which determine the strength of that substrate are exceeded by applied force. Stereotypically the footfall of modern humans and associated pressure path follows a simple pattern, although the variation on this pattern is perhaps more marked than previously thought (Bates et al. 2013a). As the heel first impacts on the ground it creates a rounded impression on a compliant substrate. This is followed by contact with the lateral side of the foot before the pressure transfer medially across the ball of the foot in the latter half of stance, ending over the first and second toe as the foot levers forwards (Elftman and Manter 1935; Morton 1935; Vereecke et al. 2003, 2005). As a consequence typically the deepest part of a footprint should occur beneath the first and second metatarsal heads which along with a deep first toe (hallucal) impression corresponds to the peak pressure at toe-off (Vereecke et al. 2003). The extent to which the lateral toes leave an impression depends on such factors as foot orientation relative to the direction of travel, the precise push-off axes and substrate properties. This simple stereotypical model assumes that plantar pressure or some measure thereof, corresponds in a simple or at least understandable fashion to depth within a given track (Bates et al. 2013b). Effectively one is considering depth an analogue for pressure.

What is not clear is the degree to which this correlation holds true in all circumstances due to the moderation of the pressure recorded by the substrate (Bates et al. 2013b). Leaving this complexity aside for the moment (see Sect. 5.3) one can vary the input of pressure and its time distribution in a number of ways. The most obvious way is to vary the speed at which an individual walks, an increase in speed should increase the force applied to the ground and may also vary the pressure distribution as the bones in the foot lock together more firmly to become a more rigid lever (e.g., Rosenbaum et al. 1994; Burnfield et al. 2004; Segal et al. 2004; Taylor et al. 2004; Warren et al. 2004; Pataky et al. 2008). One can vary the limb properties, chiefly the femur length, and the flexibility of the pelvis and trunk (Levine et al. 2012). To a certain extent this will vary with individual body proportions and pathologies, but is also particularly relevant when examining extinct human species (Vereecke et al. 2005). The centre of mass of an individual may vary and as weight increases which tends to add mass at the front/anterior and therefore impacts on balance and potentially the distribution of pressure through the various stages of stance. The behaviour of the individual may also be relevant, for example their eye gaze, and body/arm orientation may cause variations in pressure as can carrying a bag or an object. There is also an assumption here that unless an individual track-maker has some type of foot pathology the behaviour of their foot is always consistent. This may not always be the case, in some people the foot may show much higher levels of midfoot mobility than is traditionally assumed reflecting midfoot dorsiflexion (mid-tarsal break) and the way the bones lock together to varying degrees in order to form a rigid lever (Bates et al. 2013a). All of these factors make the distribution of plantar pressure for an individual a highly distinct feature, varying to different degrees from the stereotypical pattern (Pataky et al. 2012; see Sect. 7.3). The level of distinction is an intriguing question and critical to understanding the degree to which variation between species can be determined. One needs to not only to understand the degree of inter-species, but also intra-species variation before one can say with any certainty whether these differences are likely to be sufficiently great enough to be revealed in different track topologies.

The other side of the problem is the degree to which the substrate (sediment) actually records gait. Effectively what does the pattern of depth across the plantar surface actually relate to and if this is plantar pressure to what extent is this moderated by sediment properties? There are two elements to this. The first is the degree to which an individual senses a substrate and modifies their gait accordingly. We have all no doubt walked on an icy or muddy surface and as our feet begin to skate loosing traction beneath us we have shortened our stride, slowed our pace, become more tentative in our footfall and subconsciously allowed the flexibility in our foot to compensate for that instability. We shift our weight and therefore pressure to retain balance or flex the toes to acquire more grip and counter any unwanted movement. We are unconsciously modifying our gait and pattern of footfall in accordance with the properties beneath our feet something which is evident when one walks bare foot on the beach and looks at the tracks produced (e.g., Lejeune et al. 1998; Ferris et al. 1999). The very act of extracting a foot from a deep impression may also modify our gait properties. The second way in which substrate impacts on the tracks created is through the properties of the sediment itself. The way in which a substrate accommodates and then holds the void created by the foot will depend on the properties of the sediment and its mobility beneath and around the foot, particularly in the natural shear zone created between the plantar surface of the foot and the base of the track. On a hard and therefore non-compliant surface the foot makes no impression, instead the soft tissue will deform around the skeletal structure of the foot. In completely soft sediment whose strength is far less than the applied pressure, the foot will just sink and continue to do so until it meets with increased resistance. In most situations the sediment consolidates and compresses or a harder substrate is encountered at depth which begins to bear the weight of the individual (Allen 1997). The depth at which this occurs is dependent on the applied force and the vertical stratigraphy of the sediment and the rate of consolidation or strength hardening that occurs. The stability of the track post-formation is also critical; is the material strong enough to withhold the vertical or semi-vertical track walls from collapsing?

The interpretation of human tracks is therefore dependent on several key questions: (1) how unique is the pressure distribution to a given track-maker; (2) what is the range of typical behaviours and patterns for any given human species or set of individuals, and what levels of variance are there around these norms; (3) to what extent does this vary with issues of body mass and behaviour; (4) to what extent can tracks from different substrates be compared; and (5) what variance is there around the sedimentological properties at a given site and how does this add to the variance between tracks in a given trackway? These are the fundamental questions which need to be addressed to interpret human tracks and we will endeavour to address some of them within this volume.

1.4 Track Resources

Throughout this book we use a series of resources to help illustrate a range of aspects. The first of these is based on unpublished track data collected in 2007 by the senior author from 254 individuals working at Bournemouth University (males N = 101; females N = 153; 97 % Caucasian; 2–62 years old with mean of 34 years). Anthropometric data (age, height and weight) were recorded for this sample along with both two-dimensional and three-dimensional tracks. Static two-dimensional tracks, using pressure sensitive paper, were taken of each subject's right foot. At least four tracks were recorded in 3D – two rights and two lefts – walking barefoot at a comfortable/natural speed along an 5 m walk-way the central three metres of which consisted of a sediment tray, 90 mm deep filled with soft damp sand. Individual tracks were photographed and scanned using a VI900 Konica-Minolta optical laser scanner. Contour maps for a series of 12 tracks from this sample are reproduced in the Appendix and are referred to at various points to help illustrate key points.

The second resource used throughout this book are two prominent, in terms of their length, trackways from a site close to Walvis Bay in Namibia and are described by Morse et al. (2013; see Sect. 3.2.3). The longest of these two trails consists of over 70 individual tracks and the local geo-tourist guide who visits the site with a line of clients astride quad-bikes each day describes the trackway as being made by 'Old Harry' on route to the delights of Walvis Bay. The trackway has a consistent step and stride length (0.656 ± 03 m) and stride length (1.386 ± 02 m) and appears to post-date most of the other tracks on the site which consist of both domesticated and wild animals and a large number of short human trackways made by individuals potentially tending and watering flocks. The value of 'Harry's Trackway' is its length and it is introduced here and used throughout the book to illustrate the application of different methods and inferences. While it is not good practice to anthropomorphise, and the gender of the track-maker is unknown, for ease of reference throughout the book we use the term 'Harry's Trackway'. About 8 m to the south is a parallel trackway, consisting of slightly smaller tracks leading the same tour guide to refer to it as 'Harriet's Trackway'. Again we use the colloquial term to identify the trackway but recognise that the gender of the track-maker is not known.

1.5 Summary

In the following chapter we review the range of methodological and analytical tools that are need to study human tracks providing the foundation for what follows. Before looking in detail at how substrate and taphonomy (Chap. 5) may modify the topology of a track and therefore the inferences that can be made from it we provide a review of World tracksites (Chap. 3) in order to give a flavour of the different types of depositional environment in which tracks are preserved and also review some of the challenges associated with their conservation (Chap. 4). In Chap. 6 we explore the inferences that can, and crucially cannot, be made from human tracks and evaluate their value within both archaeology and palaeoanthropology. Based on this Chap. 7 looks at how the study of fossil tracks may help forensic scientists in the study of trace evidence at crime scenes, in the form of footwear and barefoot impression, before we conclude with a brief chapter outlining what we see as the future research agenda for human track studies. This is just one of many ways in which this material could be organised and the book is not necessarily meant to read in linear order, but we do encourage the reader to first look at the methods in Chap. 2 before browsing at their leisure through the later chapters.

References

Agnew N, Demas M (1998) Preserving the Laetoli footprints. Sci Am 279:44–55

Aldhouse-Green SHR, Whittle AWR, Allen JRL et al (1993) Prehistoric human footprints from the Severn Estuary at Uskmouth and Magor Pill, Gwent, Wales. Archaeologia Cambrensis 141:14–55

Allen JRL (1997) Subfossil mammalian tracks (Flandrian) in the Severn Estuary, S.W. Britain: mechanics of formation, preservation and distribution. Philos Trans R Soc Lond Ser B352:481–518

Avanzini M, Mietto P, Panarello A et al (2008) The Devil's trails: Middle Pleistocene human footprints preserved in a volcaniclastic deposit of Southern Italy. Ichnos 15:179–189

Bates KT, Collins D, Savage R et al (2013a) The evolution of compliance in the human lateral mid-foot. Proc R Soc Lond Ser B280. doi:10.1098/rspb.2013.1818

Bates KT, Savage R, Pataky TC et al (2013b) Does footprint depth correlate with foot motion and pressure? J R Soc Interface 10:1742–5662. doi:10.1098/rsif.2013.0009

Bennett MR, Harris JWK, Richmond BG et al (2009) Early Hominin foot morphology based on 1.5 million year old footprints from Ileret, Kenya. Science 323:1197–1201

Bennett MR, Falkingham P, Morse SA et al (2013) Preserving the impossible: conservation of soft-sediment hominin footprint sites and strategies for three-dimensional digital data capture. PLoS One 8:e60755

Bodziak WJ (2000) Footwear impression evidence: detection, recovery, and examination. CRC Press, Boca Raton

Bramble DM, Lieberman DE (2004) Endurance running and the evolution of *Homo*. Nature 432:345–352

Brown T (1999) The science and art of tracking – nature's path to spiritual discovery. Berkley Books, New York

Burnfield JM, Few CD, Mohamed OS et al (2004) The influence of walking speed and footwear on plantar pressures in older adults. Clin Biomech 19:78–84

Crompton RH, Pataky TC, Savage R et al (2012) Human-like external function of the foot, and fully upright gait, confirmed in the 3.66 million year old Laetoli hominin footprints by topographic statistics, experimental footprint-formation and computer simulation. J R Soc Interface 9:707–719

Deino AL (2011) $^{40}Ar/^{39}Ar$ dating of Laetoli, Tanzania. In: Harrison T (ed) Paleontology and geology of Laetoli: human evolution in context. Springer, Dordrecht, pp 77–97. doi:10.1007/978-90-481-9956-3-4

Dingwall HL, Hatala KG, Wunderlich RE et al (2013) Hominin stature, body mass, and walking speed estimates based on 1.5 million-year-old fossil footprints at Ileret, Kenya. J Hum Evol. doi:10.1016/j.jhevol.2013.02.004
Donovan SK (1994) The palaeobiology of trace fossils. Wiley, Chichester
Elftman H, Manter J (1935) The evolution of the human foot, with especial reference to the joints. J Anat 70:56
Ferris DP, Liang K, Farley CT (1999) Runners adjust leg stiffness for their first step on a new running surface. J Biomech 32:787–794
Fornós JJ, Bromley RG, Clemmensen LB et al (2002) Tracks and trackways of *Myotragus balearicus* bate (Artiodactyla, Caprinae) in Pleistocene aeolianites from Mallorca (Balearic Islands, Western Mediterranean). Palaeogeogr Palaeoclimatol Palaeoecol 180:277–313
Hunt AP, Lucas SG (2007) Tetrapod ichnofacies: a new paradigm. Ichnos 14:59–68
Kim JY, Kim KS, Lockley M et al (2008) Hominid ichnotaxonomy: an exploration of a neglected discipline. Ichnos 15:126–139
Kim KS, Kim JY, Kim SH et al (2009) Preliminary report on hominid and other vertebrate footprints from the Late Quaternary strata of Jeju Island, Korea. Ichnos 16:1–11
Leakey MD, Harris JM (1987) Laetoli: a Pliocene site in northern Tanzania. Clarendon, Oxford
Leakey MD, Hay RL (1979) Pliocene footprints in the Laetoli beds at Laetoli, northern Tanzania. Nature 278:317
Lejeune TM, Willems PA, Heglund NC (1998) Mechanics and energetics of human locomotion on sand. J Exp Biol 201(13):2071–2080
Leonardi G (1987) Glossary and manual of tetrapod footprint palaeoichnology. Publicação do Departemento Nacional da Produção Mineral Brasil, Brasília
Levine D, Richards J, Whittle MW (2012) Whittle's gait analysis, 5th edn. Elsevier Health Sciences, London
Lockley MG (1991) Tracking dinosaurs: a new look at an ancient world. Cambridge University Press, Cambridge
Lockley MG, Hunt AP (1995) Dinosaur tracks and other fossil footprints of the Western United States. Columbia University Press, New York
Lockley MG, Meyer CA (2000) Dinosaur tracks and other fossil footprints of Europe. Columbia University Press, New York
Lockley MG, Roberts G, Kim JY (2008) In the footprints of our ancestors: an overview of the hominid track record. Ichnos 15:106–125
Manning PL (2004) A new approach to the analysis and interpretation of tracks: examples from the Dinosauria. In: McIlroy D (ed) The application of ichnology to palaeoenvironmental and stratigraphical analysis. Geological Society of London, Special Publication 228:93–128
Marty D, Strasser A, Meyer CA (2009) Formation and taphonomy of human footprints in microbial mats of present-day tidal-flat environments: implications for the study of fossil footprints. Ichnos 16:127–142
Meldrum DJ, Lockley MG, Lucas SG et al (2011) Ichnotaxonomy of the Laetoli trackways: the earliest hominin footprints. J Afr Earth Sci 60:1–12
Morse SA, Bennett MR, Liutkus-Pierce C et al (2013) Holocene footprints in Namibia: the influence of substrate on footprint variability. Am J Phys Anthropol. doi:10.1002/ajpa.22276
Morton DJ (1935) The human foot. Columbia University Press, New York
Pataky TC, Caravaggi P, Savage R et al (2008) New insights into the plantar pressure correlates of walking speed using pedobarographic statistical parametric mapping (pSPM). J Biomech 41:1987–1994
Pataky TC, Mu T, Bosch K et al (2012) Gait recognition: highly unique dynamic plantar pressure patterns among 104 individuals. J R Soc Interface 9:790–800
Raichlen DA, Gordon AD, Harcourt-Smith WEH (2010) Laetoli footprints preserve earliest direct evidence of human-like bipedal biomechanics. PLoS One 5:e9769. doi:10.1371/journal.pone.0009769
Robbins LM (1985) Footprints: collection, analysis, and interpretation. CC Thomas

Roberts G, Gonzalez S, Huddart D (1996) Inter-tidal Holocene footprints and their archaeological significance. Antiquity 70:647–651
Rosenbaum D, Hautmann S, Gold M, Claes L (1994) Effects of walking speed on plantar pressure patterns and hindfoot angular motion. Gait Posture 2:191–197
Schmincke H-U, Kutterolf S, Perez W et al (2009) Walking through volcanic mud: the 2,100 year old Acahualinca footprints (Nicaragua). Bull Volcano 71:479–493
Schmincke HU, Rausch J, Kutterolf S et al (2010) Walking through volcanic mud: the 2,100 year-old Acahualinca footprints (Nicaragua) II: the Acahualinca people, environmental conditions and motivation. Int J Earth Sci 99(1):279–292
Segal A, Rohr E, Orendurff M et al (2004) The effect of walking speed on peak plantar pressure. Foot Ankle Int 25:926–933
Taylor AJ, Menz HB, Keenan AM (2004) Effects of experimentally induced plantar insensitivity on forces and pressures under the foot during normal walking. Gait Posture 20:232–237
Thulborn T (1990) Dinosaur tracks. Chapman and Hall, London
Thulborn T (2012) Impact of sauropod dinosaurs on lagoonal substrates in the Broome Sandstone (Lower Cretaceous), Western Australia. PLoS One 7(5). doi:10.1371/journal.pone.0036208
Vereecke E, D'Août K, De Clercq D et al (2003) Dynamic plantar pressure distribution during terrestrial locomotion of bonobos (*Pan paniscus*). Am J Phys Anthropol 120:373–383
Vereecke E, D'Août K, Van Elsacker L et al (2005) Functional analysis of the gibbon foot during terrestrial bipedal walking: plantar pressure distributions and three-dimensional ground reaction forces. Am J Phys Anthropol 128:659–669
Warren GL, Maher RM, Higbie EJ (2004) Temporal patterns of plantar pressures and lower-leg muscle activity during walking: effect of speed. Gait Posture 19:91–100
White TD, Suwa G (1987) Hominid footprints at Laetoli: facts and interpretations. Am J Phys Anthropol 72:485–514

Chapter 2
Methods of Data Capture and Analysis

Abstract The sophistication and quality of field data obtained from human tracksites has increased dramatically during the last decade from the largely descriptive papers of Holocene tracksites common before the late 1990s to the more sophisticated data-rich papers of recent years. There are exceptions of course to this generalisation largely around the tracks at Laetoli which drove early innovation in methods. In this chapter we review the methods and approaches that can be adopted at human tracksites and equip the interested researcher with the knowledge necessary to execute such investigations themselves given suitable excavation permits and permissions. We recognise four broad stages to the process each of which is considered in turn: (1) geo-prospection and excavation; (2) recognition of human tracks and their dating; (3) methods of digital data capture; and (4) methods of analysis.

2.1 Geo-prospection and Excavation

2.1.1 Finding Tracks

One has to first find the tracks. Within the geological record, assemblages of vertebrate tracks are found in a wide range of depositional settings typically characterised by fine-grained sediment and a depositional regime that allows for rapid burial. Lacustrine or marine quiet water environments with rapid variation in lake or sea level that cause rapid shoreline transgressions are ideal (e.g., Melchor et al. 2006; de Gilbert and Sáez 2009; Krapovickas et al. 2009). Human tracks are found in similar depositional systems, although due to the high profile of such as sites Laetoli they are often perceived to be mainly associated with volcanic ash, when in fact most records pertain to cave sites (Lockley et al. 2008; Table 3.1). Figure 2.1 provides a schematic model of the range of human track-forming environments documented so far in the literature and consequently, given appropriate ages, those that might contain tracks. Ideal conditions for the preservation of human tracks can be summarised as follows:

1. The presence of track-makers in sufficient numbers to leave a density of tracks such that they stand a chance of being preserved and later discovered, preferably in association with only a few other vertebrates so that the tracks are not lost to trampling (Laporte and Behrensmeyer 1980). A point of congregation, or a

M.R. Bennett, S.A. Morse, *Human Footprints: Fossilised Locomotion?*,
DOI 10.1007/978-3-319-08572-2_2

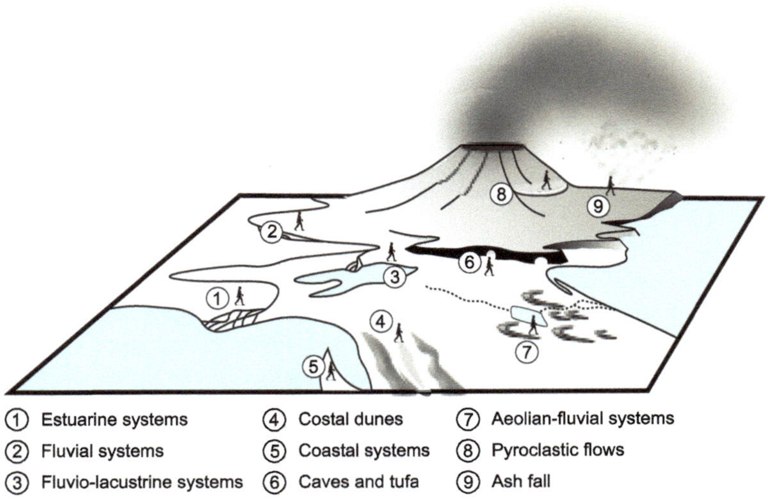

Fig. 2.1 Conceptual model of the landscape and depositional systems in which human tracks have been found to date. Compare this to Table 3.1 which lists some of the main human track sites around the World

tracking surface of restricted area, may help this process by increasing the density of tracks and also by restricting the search area, however equally this may lead to many tracks per unit area and associated overprinting and poor preservation. Human tracksites can in general be separated out into 'congregation sites' and 'transit sites'. The latter occur along a trail or pathway, while former are often associated with some form of water source such as a lake or pool or a source of habitation in the case of a cave.

2. A fine-grained substrate with a consistency able to take, and hold the impression as well as cast its detail.
3. A hardening process and/or rapid burial of the track. Examples of hardening include the lithification of volcanic ash, desiccation causing sediments to bake and/or become cemented through the concentrations of salts precipitated from groundwater.
4. Limited post-depositional compression and/or deformation of the track bearing horizon with just sufficient subsequent erosion to exhume or partially exhume the horizon and preferably only once.
5. Finally an observer – in the right place, at the right time – able to recognise the track in vertical cross-section or recognise the potential for tracks on a partially exposed or near-surface layer.

The recognition of tracks in vertical cross-section is important to this process and many tracks go undetected as a consequence of a lack of field recognition. Van der Lingen and Andrews (1969) describe the structures formed by 'hoof-prints' in beach sand, a theme developed further in the context of other animal tracks by a

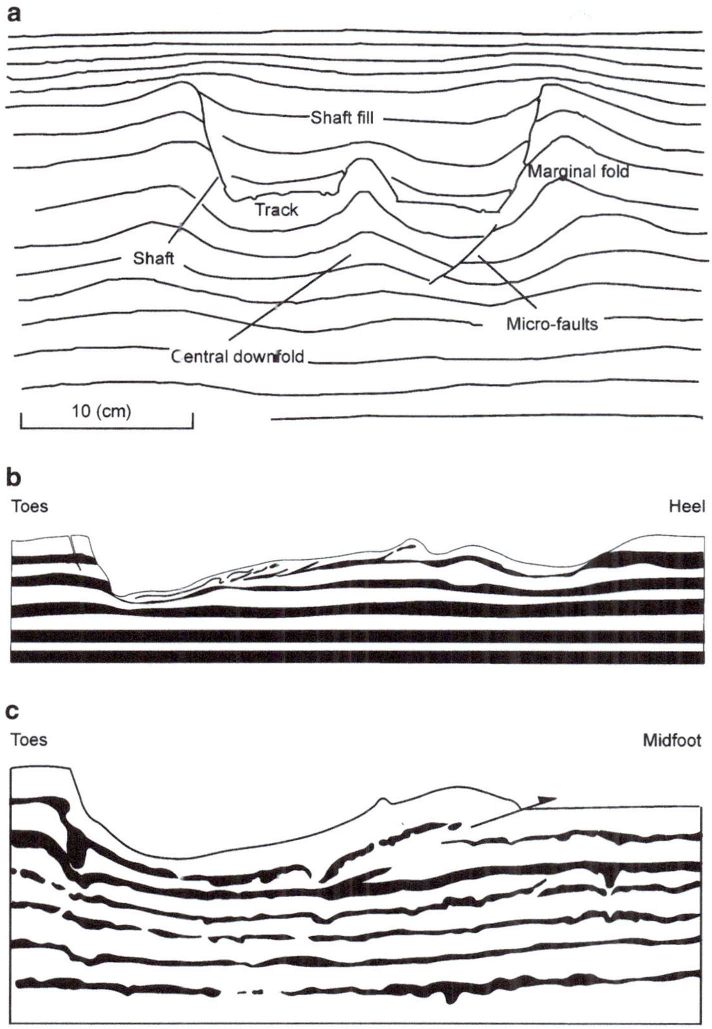

Fig. 2.2 Cross-sections through tracks. (**a**) Typical animal track following the work of Lea (1996). (**b**) Cross-section of a human track produced in a sand box by the authors using coloured sand in alternate layers. (**c**) Sketch traced from a photograph in Brown (1999) of a cross-section through the forefoot of a human track. In this case the layers are produced by alternating layers of sand/earth and flour

number of authors (e.g., Laury 1980; Allen 1989, 1997; Loope 1986; Scrivner and Bottjer 1986; Lea 1996). According to Lea (1996) the features associated with the ideal track in cross-section include (Fig. 2.2a): (1) a steep shaft which typically truncates adjacent beds; (2) a track at the base of the shaft representing the plantar surface of the animal's foot; (3) a shaft fill which may be sedimentologically distinct from the host bed; and (4) a deformed zone surrounding the shaft which may include

Fig. 2.3 Cross-section through two in filled hippopotami tracks at Kobi Fora (GaJi10) in northern Kenya

a central down-fold and marginal up-folds. Figure 2.3 shows the track of a hippopotami in the side wall of an excavation in Kenya and shows some of the features in Fig 2.2. Human tracks are often much shallower than those of other animals and their cross-sectional morphology has not been documented in detail. Figure 2.2b and c show cross-sections through modern human tracks, focusing on the under tracks. The essential features described above are present although the level of marginal deformation evident is dependent on the substrate and pattern of strain which has taken place (see Sect. 5.2). The challenge in recognising tracks in cross-section is to distinguish them from other bedding surface disturbances caused by such things as piping subsidence or soft-sediment deformation.

2.1.2 Excavation and Mapping

Excavation may be required to uncover a palaeosurface and individual tracks may need to be exhumed carefully to reveal the true track. In some cases the sediment infill may be dry allowing the print to be cleaned with a brush, in other cases indurated sediment may need to be removed delicately using a dental pick. It is important to ensure that tracks are exhumed carefully and that the features shown are in fact true. This involves a working knowledge of sedimentology and a careful, patient eye. The reader is directed to the unpleasant speculation on the part of some researchers about the damage done to the Laetoli tracks during their excavation (Tuttle et al. 1990). Once exposed a palaeosurface may be mapped in a variety of different ways. At its simplest larger areas can be gridded and sketched using quadrats or simple tape and offset surveys (Fig. 2.4a). Other ichnologists favour placing paper or acetate sheets over the tracks and tracing outlines directly.

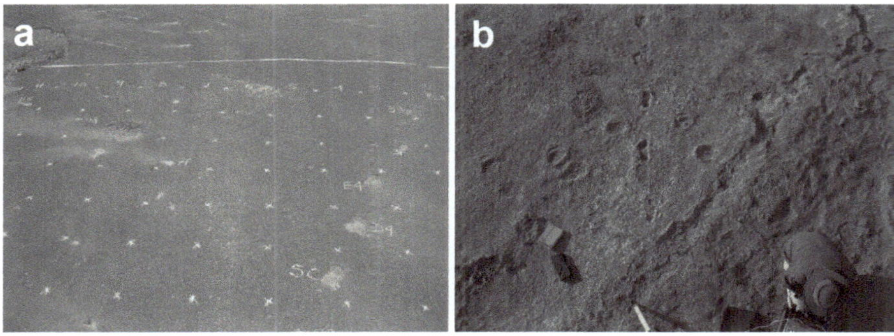

Fig. 2.4 Mapping footprint surfaces. (**a**) Chalk grid set-out to map a potential ichnological surface in Central Mexico. (**b**) View from mast-cam showing part of Harry's Trackway at Walvis Bay (Namibia; Sects. 1.4 and 3.2.3). This 5 m high mast had an automatic camera on a powered gimbal allowing it to pan and rotate with a direct video feed and capture facility on the attached laptop. The system worked well until wind-blown dust destroyed the camera lens

More sophisticated approaches for smaller areas now employ the use of near-surface aerial photographs either from a step ladder or a mast-mounted camera (Fig. 2.4b). The use of surveyed ground control points – fluorescent coloured wooden cubes (1 cm) held in place either by 'white-tack' or attached to the heads of 6-in. nails in the case of beach experiments – allows these low level oblique or semi-vertical photographs to be geo-rectified in appropriate image or mapping software. Traditionally big A-frames have also been used to mount conventional cameras and take stereo images of excavations (Breithaupt et al. 2004).

2.2 Recognising Human Tracks

The correct recognition of human tracks is important and not an issue without some controversy. For example Leakey (1978) tentatively suggested that five prints at Laetoli Site-A, may have been made by a bipedal hominin. This interpretation was supported according to Tuttle (2008) by some authorities, but was after some debate rejected and the tracks tentatively re-interpreted as those of a small facultative bipedal Pliocene bear (Tuttle 1984, 1987). Prints with poor anatomical topology which don't form clear trackways may also challenge interpretation, for example such as those preserved in peat at Kenfig on the coast of South Wales (Bennett et al. 2010), or the recent prints from Happisburgh on the UK Norfolk coast (Ashton et al. 2014). In order to resolve the debate around the controversial (González et al. 2006; Feinberg et al. 2009) potential tracks at Valsequillo, Mexico Morse et al. (2010) devised a test based around two self-evident criteria. Firstly, a print must reflect the basic anatomy and function of a human foot in terms of dimensions, anatomical proportions and plantar pressure patterns as measured by variations in track shape and depth. Something which is of course mediated through such things as substrate properties, and kinematic variables such as the speed of walking, as well as allowing for potential evolutionary changes in both foot anatomy and locomotor style. Secondly, by definition tracks should form part of a trackway. However it is important to acknowledge that some tracksites contain isolated prints either where there is a palimpsest of superimposed animal prints or clear variations in substrate properties such as water content. In responding to the first of these criteria Morse et al. (2010) argued that the pattern of depth distribution within a footprint was distinctive, reflecting the stereotypical distribution of plantar pressure shown in Fig. 1.2b and that this was likely to hold across most species from *Australopithecus* through *Homo* genera. They devised a system of geometrically placed landmarks that could be applied consistently irrespective of changes in anatomy, substrate and track definition. They enclosed a track within an ellipse and placed landmarks where the footprint boundary intersects the primary axes of the ellipse. In this way it is possible to place four landmarks on a print irrespective of their anatomical definition and to then locate a further landmark to denote the deepest point in the track (Fig. 2.5). Using tracks from a range of geographical locations, environments and hominin species they were able to show that the deepest point of a track plotted either in one or other of the lateral hemispheres (heel or metatarsal areas) of the enclosing ellipse rather than in the centre.

In the case of the potential Valsequillo tracks not only did maximum depth plot in the middle of the ellipse but the prints were much broader than the reference tracks. Added to which the absence of trackways clinched the case and lead to their rejection as human tacks (Morse et al. 2010). In contrast in the case of Kenfig (Bennett et al. 2010) the depth distribution and widths were consistent with the reference tracks even though the individual tracks had poor anatomical definition. In addition a single clear right-left trackway assisted this process. The capture of good three-dimensional digital elevation data on a track is crucial to their correct identification and the recognition of tracks as being human deserves real care especially where their interpretation is of potential significance.

2.3 Dating Human Tracks

There are two aspects to this, firstly the relative sequence of events at a tracksite and secondly the absolute age of those events. The relative age can be determined by consistent cross-cutting patterns of individual tracks and/or trackways and is particularly important at densely tracked sites. The famous Laetoli trackways provide one example, where the G-2 and G-3 trackways are superimposed with the smaller G3-Tracks being superimposed on to those of G-2 (see Sect. 3.1.1). Relative track chronology is also relevant in a forensic context where both two- and three-dimensional tracks may provide evidence to help corroborate witness testimony with respect for example to the sequence of movements that took place at a crime scene (Fig. 2.6).

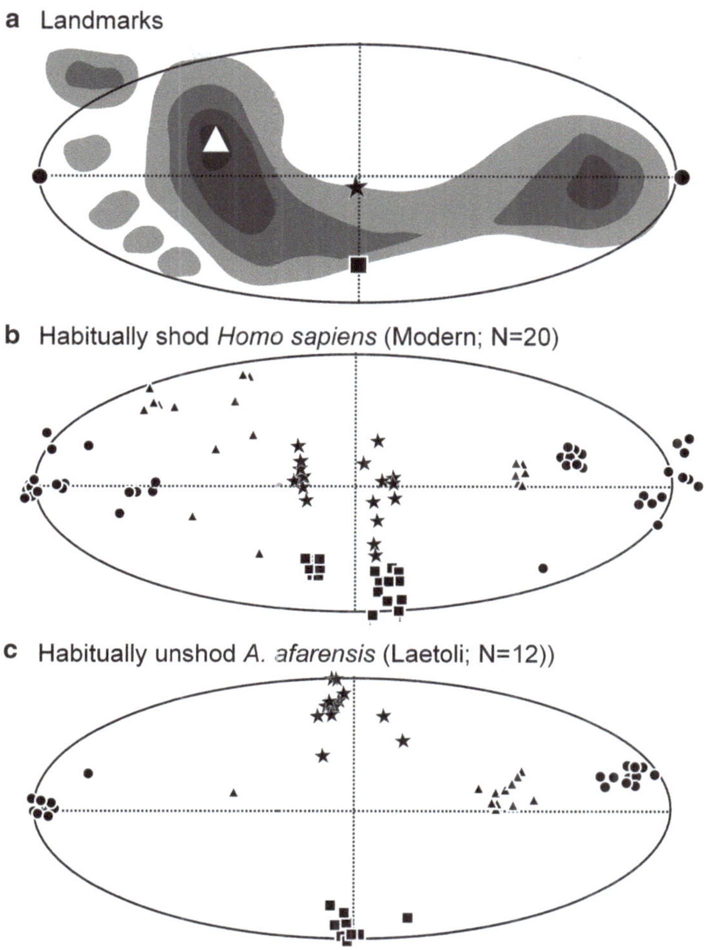

Fig. 2.5 Footprint recognition after Morse et al. (2010). (**a**) Shows the placement of four geometrical landmarks. The track is enclosed within an ellipse the axes of which are used to place the landmarks. The critical landmark is the deepest point represented by the *triangle*. (**b**) The pattern of landmarks for a series of modern tracks (N = 20; *Homo sapiens*). Note the landmark – *triangles* – marking the deepest points plot either in the left or the right hemispheres. (**c**) The pattern of landmarks for a series of tracks from the G-1 Trackway at Laetoli in Tanzania (N = 12. *Australopithecus afarensis*). Note that again the deepest points plot in either the right of left hemispheres. In all cases two-dimensional co-ordinates were obtained for each landmark from footprint scans and these were subject to a Generalised Procrustes Analysis within PAST

The absolute age of a series of fossil tracks is of often critical in determining the archaeological or palaeoanthropological significance of a human tracksite. It is also something that can on occasions be less than straightforward attracting considerable controversy and debate. Our intention here is not to review the range of possible dating options which one could apply, which depend on the age and nature of the

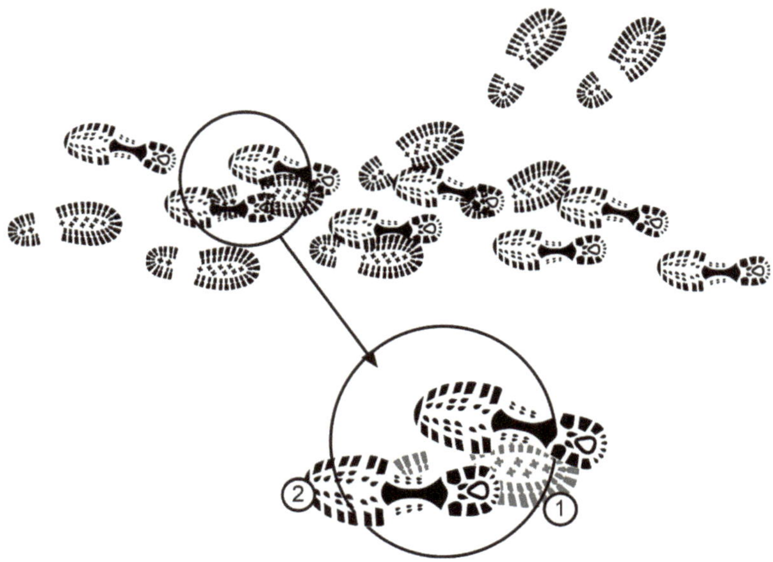

Fig. 2.6 Cross-cutting tracks. Conceptual illustration of the ways in which cross-cutting relationships can be used to work out the relative order of two trackways

deposit in question (Fig. 2.7a), but instead to explore the particular stratigraphic challenges which exist at some fossil tracksites. Dating of tracks can be undertaken in one of three ways: (1) by their assumed association with dated archaeological artefacts and periods of known occupation found near, or at a tracksite; (2) by lithostratigraphic correlation of the beds containing the tracks, or bounding units, with those of known age elsewhere; and (3) by direct dating of the sediment, and/or elements within it, that either contain the tracks or occur above or below them. Many of the European cave sites containing human tracks are dated by the association with known period of occupation (Lockley et al. 2008), similarly the recently discovered tracks at Happisburgh on the Norfolk Coast (UK) are found in beds which can be correlated on both lithostratigraphic and biostratigraphic grounds with those dated elsewhere along the coast (Ashton et al. 2014). The tracks at Walvis Bay illustrate the need for caution in some types of environment, however (Fig. 2.8). Here periodically flood waters from the usually dry Kuiseb River escape their channel and flow between networks of large sand dunes leaving horizontal silt and clay sheets on which human and animal footprints are preserved (Kinahan 1996). The dunes migrate over these surfaces revealing tracks in inter-dune slacks, but while these surface are lithostratigraphically similar they are in fact diachronous, ranging in age from as little as 500 to over 1,500 years old (Morse et al. 2013; Fig. 2.8). Event stratigraphy may also play a role in dating tracksites. For example, the tracks at Ileret and Koobi Fora in northern Kenya are dated to approximately 1.5 and 1.4 Ma respectively on the basis of beds of volcanic ash (tephra/tuff) found above and below the track horizons, which can be correlated geochemically to tephra

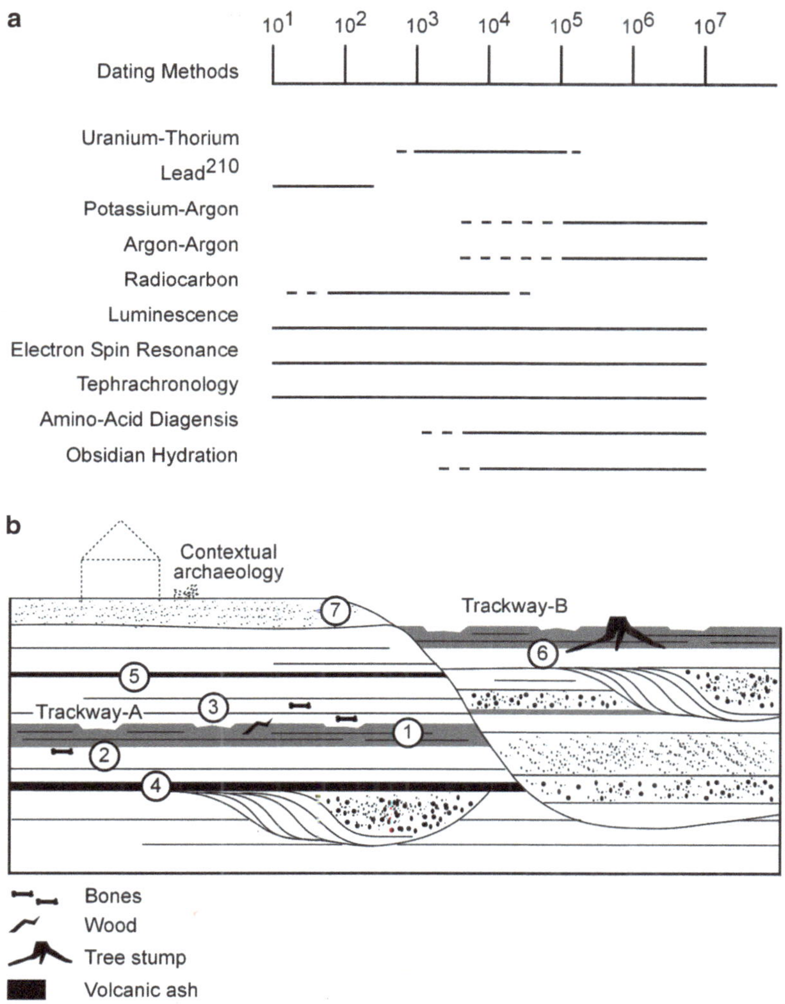

Fig. 2.7 Dating of fossil tracks. (**a**) Time range for typical dating techniques applied to tracks and their bounding deposits. (**b**) Schematic model of two dating scenarios, see text for detailed explanation

samples that have been dated through Ar/Ar techniques (McDougal and Brown 2006; Bennett et al. 2009). In particular the tracks at Ileret are dated with respect to three volcanic ash layers and assumptions about the rate of sedimentation are required to refine this broad geochronology; the upper and lower footprint horizons are assumed to be approximately 10 k years apart in age on this basis. In another case the tephra containing the human tracks at Acahualinca has been correlated with the dated eruptive history from the source volcano (Schmincke et al. 2009), or that at Avellino below Vesuvius in Italy which can be dated precisely to a specific phase of the Plinian eruption in 3.9 ka BP (Mastrolorenzo et al. 2006).

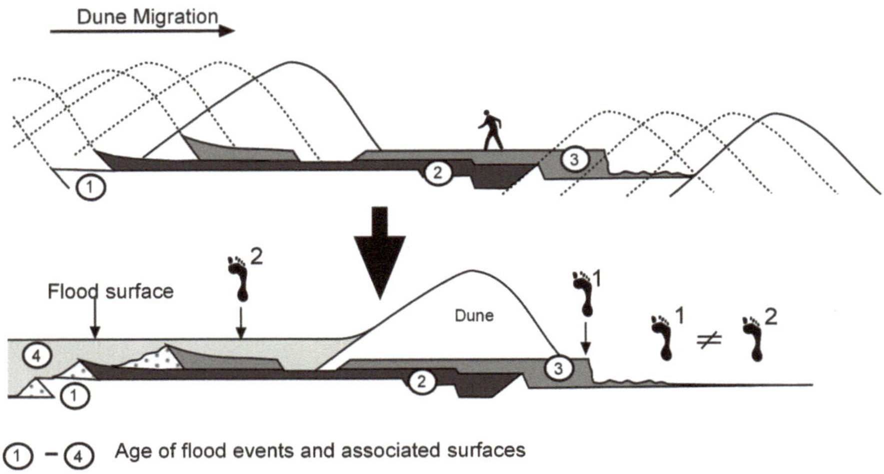

Fig. 2.8 Diachronous nature of human and animal track surface south of Walvis bay in the Namib dune field

Where direct dating of tracks is undertaken it is important to be clear what it is that is being dated; the episode of track-making, or the deposit in which the tracks are preserved? The two need not be the same. Take for example a fine-grained mud layer, freshly deposited, damp and pristine, in fact a perfect blank canvass on which to walk. If a line of tracks is duly recorded by some willing subject and left exposed they will begin the process of taphonomy, perhaps drying and hardening under the sun. Rainfall or a brief flood event may re-activate the mud surface at any time and a second track-maker may leave their mark also. This process will continue until the tracked surface is buried and sealed beneath a layer of capping sediment. The track-making episode may last several hours, days or months and there is also a chance that the surface could be re-activated if re-exposed by erosion and the sediment beneath remains soft. In this case dating of the deposit need not necessarily provide an accurate date of the episode of track-making. In most cases the relationship between the two is clear, but this may not always be true.

We can explore some of these issues further with respect to the hypothetical examples in Fig. 2.7b. Here we have two tracked surfaces one buried deeply within the sedimentary succession (Trackway-A) and a second exposed at the surface (Trackway-B) both formed in fluvial muds for sake of argument. In the case of Trackway-A, we have a range of possible dating options which are numbered 1–5; ideally we are looking to bracket the track-bearing horizon with dates below and immediately above the tracks. For example we may be able to extract an Optically Stimulated Luminescence data for arguments sake from the overlying sands(3), perhaps one from the layer below the track horizon(2), or alternatively a radiocarbon date from the organic flotsam within the tracked layer(1), notwithstanding the fact that this flotsam could well be much older than the deposit itself. Alternatively we may use the two volcanic ashes within the succession (Fig. 2.7b, numbers 4 and 5) to bracket the block of strata in which the tracks occur. There are lots of options assuming that datable materials are present, but simply providing a date for the track bearing horizon is potentially not sufficient to constrain its age. Ideally one should try to obtain a date from the overlying deposits (i.e. post-track) and compare this with the age of the track horizon or the beds immediately below it. Assuming that the dating, preferably from multiple samples and/or methods to allow internal corroboration, is successful and in turn the results are consistent with any available external corroboration from other sites, then we will ideally have an age range during which track-formation took place. Note that contrary to many presentations of age within the literature on human tracks we are not able to provide a precise date, but are simply able to bracket a range of possible ages. It is important to stress that without good stratigraphic context absolute dates may be interpreted in different ways. The human tracks of Jeju Island in South Korea provide an excellent example of this, with some workers ascribing their age to the late Pleistocene (~19–25 ka), while others believe that they may be as young as 3.7 ka (cf. Cho et al. 2005; Kim et al. 2010). The issue here is the interpretation of the complex and laterally variable stratigraphy of the island which is associated with numerous eruption cones and re-worked volcaniclastic deposits (Sonh et al. 2002; Sohn et al. 2012). Despite the eleven radiocarbon dates presented in Kim et al. (2010) they do not demonstrate to the satisfaction of some workers, either on the basis of lithology or geochemistry, that the deposits they date are in fact the same as those that bear the footprints. Until there is an agreed consensus with respect to the stratigraphic context of the track bearing horizons it is impossible to interpret the range of available dates correctly and without challenge.

The case of Trackway-B (Fig. 2.7b) is more complex, with the tracks exposed at the surface. The task here is to first establish that they are indeed fossils by tracing the horizon (+tracks), through excavation if necessary, beneath adjacent in situ units. This is particularly relevant to the tracksites around the coast of the UK which outcrop on the beach after periods of erosion and a tapestry of new and old prints may be visible (e.g., Roberts et al. 1996; Bennett et al. 2010). In such cases the track-bearing horizon is exposed as part of the wave cut platform on which the current beach rests and is protected normally by that beach. The track bearing horizon has to be traced inshore where in situ beds can be seen to overlay it and a selection of tracks, a task that is not always easy since whatever remains of the current beach usually obscures this contact. The possibility of multiple episodes of exhumation of the tracked surface during historic low stands of the beache is a real possibility at such sites. The issues are illustrated by a Victorian Penny that was found by one of the authors embedded in a track exposed on the Sefton Coast by a period of beach erosion. Short of time travel there are two possible explanations for the presence of the penny, either: (1) during a previous episode of exhumation of the fossil track, the penny was washed into the track and embedded into the softened surface before it was reburied beneath the beach; or (2) a previous exhumation of the footprint surface softened it sufficiently for it to take a new (Victorian?) track in which the penny was either trampled underfoot or washed into the track as it was re-buried beneath the beach. While the former is the more likely scenario, given the consistency of the track bearing layer, the latter cannot however be completely discounted. These types of issue become even more acute in the case of cave deposits where the tracks may remain completely uncovered in remote cave passages that receive little outside detritus. They are identified as fossils primarily because they are barefoot tracks and modern cavers rarely pursue their sport without footwear! But here the potential for the surface to represent a time averaged (potentially over a significant time interval in this case) record of footfall is considerable so even if the age of the deposit can be determined, the age of the individual tracks may not. In the Jaguar Caves of Tennessee (Willey et al. 2005) for example, the dating is based on a surface scattering of charcoal believed to be derived from the burning torches carried by the prehistoric cavers who left the tracks. While this is a reasonable assumption there is considerable latitude here for error, especially since only a few charcoal fragments have been dated. Dating multiple fragments might give an indication as to the frequency with which the site was visited, as would detailed analysis of any cross-cutting track patterns. Human tracks from many of the European caves are simply dated by association with the archaeological evidence of cave occupation (Lockley et al. 2008).

In the hypothetical example of Trackway-B (Fig. 2.7b) we have also depicted a fossil tree stump within the footprint layer, which depending on its age, is datable by radiocarbon and while it may help to constrain the age of the deposit it is not clear how this would relate to the track-making episode. While perhaps a slightly contrived example it does actually pertain to a real one and the challenges faced when dating the human tracks in coastal peat at Kenfig in the UK (Bennett et al. 2010). The point that we are making here is that human tracks are not always easy to date and it is essential that the geoarchaeologist is aware of the potential pitfalls of doing so. We draw particular attention to the need for the following information to be sought, considered and reported in presenting age estimates for human tracks:

1. The footprints need to be first established as fossil tracks through their stratigraphic context where that is possible.
2. The depositional history of the track bearing strata, along with the mode of track formation and the subsequent taphonomy of the tracks needs to be established. Where it is possible investigators should attempt to estimate a potential 'exposure time' for the surface (e.g., hours, days, weeks or months). The frequency of burial events such as ash fall or flood episodes may help here. In addition, a site needs to be carefully investigated for evidence of potential reactivation and/or natural exhumation of the surface during its burial history and we encourage investigators to keep an open mind to this possibility even where there is no direct evidence.
3. Where at all possible dates need to be sought which bracket – above and below – the tracked surface and simply presenting a date for the tracked unit strictly speaking only provides a 'younger than' age for the tracks. Due to potential re-activation of the surface, the tracks may be significantly younger.

2.4 Methods of Digital Data Capture

Traditional methods of data capture involve detailed field-based measurements/observations, grid-based mapping as described above and some form of casting either with latex or in its crudest form via dental plaster. The importance of obtaining accurate three-dimensional data is widely acknowledged and traditional photogrammetrical approaches based on hard-copy, vertical overlapping images was pioneered at Laetoli to generate contour maps (Day and Wickens 1980; Leakey and Harris 1987). Breithaupt et al. (2004) reviewed a range of different approaches to collecting this type of data for dinosaur track sites. The issue of obtaining high quality three-dimensional data for tracks has, however, been revolutionised in recent years by the advent of small portable, high-resolution optical laser scanners and by digital photogrammetry and there are now a range of different methodological approaches available to the collection of digital track data (Bennett et al. 2013).

González et al. (2006) used a close-quarter optical laser scanner at the disputed Valsequillo tracksite in Mexico and similar technology has been used at other sites including at Ileret (Bennett et al. 2009). This involved deploying a scanner mounted on various rigs, the most sophisticated of which had a lightweight carbon fibre frame (Figs. 2.9 and 2.10). Laboratory based scans of casts of the Laetoli footprints have been used widely in a number of analyses (Raichlen et al. 2010; Meldrum et al. 2011; Crompton et al. 2012). In recent years optical laser scanning has been increasingly challenged by the availability, and increased accuracy, of soft-copy photogrammetrical software that allows digital elevation models to be generated relatively easily from multiple oblique images (Falkingham 2012; Falkingham et al. 2014). Bennett et al. (2013) provide a comparison of methods with respect to human tracks using both tracks created in the laboratory and a trackway generated on a

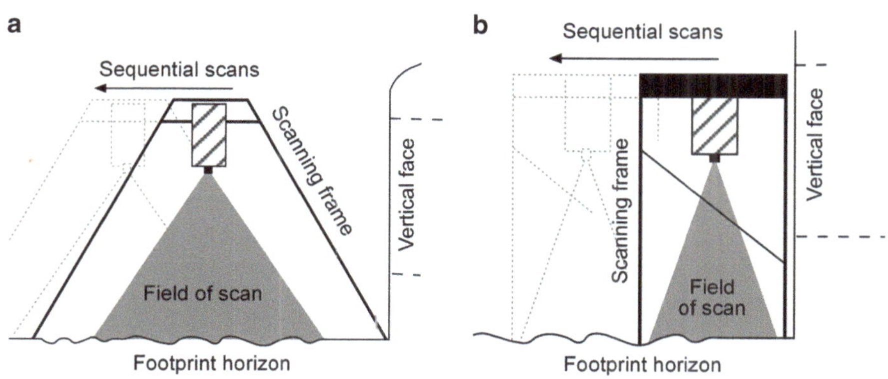

Fig. 2.9 Two alternative rigs developed by the authors to mount a VI900 Konica Minolta optical laser scanner in the field. (**a**) Triangular rig allowing the scanner to move on a central aluminium beam, ideal for taking sequential swaths along a trackway. (**b**) A compact frame made from carbon fibre developed specifically for working in the confined area of an excavation

Fig. 2.10 Optical laser scanner in the field protected by a canvas cover around a square frame made of carbon fibre as shown in Fig. 2.9b. The image shows the authors scanning human tracks at Ileret in 2009

local beach. Very little difference was observed between the two methods suggesting that both produce comparable results, although the quality of the results obtained by photogrammetry is dependent on the photogrammetrical software used with the best results currently being obtained from freeware rather than proprietary software (Falkingham 2012). The approach needs good variation in pixel textures and surface moisture can limit the quality of results obtained by photogrammetry and in some cases models may fail to build completely (Bennett et al. 2013; Ashton et al. 2014). However the principle issue with photogrammetric models is that they must be scaled either during construction or subsequently in a three-dimensional editing tool and this can limit the accuracy and consistency of measurements between models. In contrast this is not an issue with most optical laser scanners which operate with levels of accuracy and precision at a sub-millimetre scale. For example, in the experiments conducted by Bennett et al. (2013) photo-models were found to consistently underestimate distances when compared to those obtained in the field or from laser scanning.

A field scientist is currently faced, therefore, with a choice between alternative technologies namely whether to use an optical laser scanner or to adopt photogrammetry. As Bennett et al. (2013) argue issues of operational deployment are crucial here and in light of their field experience they summarise the decision-making process (Table 2.1). Photogrammetry offers advantages in the field of being easily deployed with relatively little investment in equipment and/or complex field logistics. A standard six to eight megapixel digital camera with a good quality lens is all that

Table 2.1 Summary of the relative merits involved in the field deployment of optical laser scanning versus photogrammetry following and modified from Bennett et al. (2013)

Issue	Photogrammetry	Optical laser scanning
Costs		
Hardware	Low field costs since a basic digital camera and memory cards are all that are required. Modest lab costs associated with provision of suitable CPU, dependent on the speed of processing required and software to be run; reducing all the time as standard computational power increases	High depending on the make and model of the scanner used. Low lab costs since no special computational power is required unless a large number of scanned images are being tessellated
Software	Zero to modest depending on the software used to generate photogrammetric models. Three-dimensional imaging software required for post processing and visualisation, both commercial and freeware options available	Variable, most expensive scanners come with basic three-dimensional imaging software required for post processing and basic visualisation while less expensive scanners often don't
Deployment		
Transport logistics	Easy – photo-scale, camera and memory cards. In some cases use of tripod mounted arms or A-frames may increase the equipment volume	Depending on scanner model and the support mechanism – tripod or frame – can be quite bulky. Provision of power supply via a converter and a generator, car battery or lithium ion battery
Electrical Requirements	Minimal, power is required for camera batteries and photo storage devices such as a laptop or PDA	Most scanners either require a generator, car battery or lithium ion battery with or without a power inverter, either to power the scanner directly, or to recharge a built in battery. Power is also required for PDA or laptop used to run the scanner
Data capture time	Approximately 5 min per print to take between 20 and 30 photographs per print; quicker times possible when using fixed point frames/tripod requiring a more limited number of images. It is possible to have multiple prints or areas being captured simultaneously with multiple photographers. Photographs can also be collected from Unmanned Aerial Vehicles (UAV) especially where large areas are involved, although this will increase the associated costs and logistics	Depends on the scanner model and resolution required but usually less than 1 min per scan. Limited to the number of scanners available to one field project

(continued)

Table 2.1 (continued)

Issue	Photogrammetry	Optical laser scanning
Post-processing time	Depends on the software being used and the number of images but post-processing time to generate the model can be up to 12 h, typically 15–45 min for a high resolution model	Depends on the tasks being performed and the degree of data cleansing and optimisation required but can be anything from a few minutes to 30 min maximum. Aligning multiple scans, especially from long range scanners with high data throughput can take considerable time (up to 24 h)
Reconnaissance operation and/or training?	Images can be captured by any operator with a digital camera and basic knowledge of what photographs are required	Requires access to equipment and basic training
Memory requirements	Can be managed by multiple data cards, field based download to laptop or PDA, or field based upload via internet connection. Data volumes are high depending on the individual pictures resolution; for example, one gigabyte for a trail of 10 prints	Depends on the make and model of scanner, some scanners can record directly to a data card, most required laptop operation. Typical file sizes are between 1 and 5 megabytes per print, though high resolution scans of large areas (e.g. whole or partial track sites) can be many Gb in size
Risks to site	Damage can be high from feet of photographer taking multiple images from different angles; damage from the feet of tripods or other fixed arm camera mounts. These can be overcome through the use of UAV's although their use increases costs and logistics	Damage from tripods or scanner frames can be high and use at sites with a high water content/in the tidal zone is dangerous to the equipment and operator
Accuracy of outcome		
Prohibitive environmental conditions	Sunlight & intense shadow can be problematic and shading may be required for the whole area of the print depending on the colour of the substrate and angle of the sun. Wind-blown dust and rain may hinder operation. Wet rock/sediment surfaces or those with residual water content can limit the accuracy of some models especially where it is variable across a surface	Most high resolution optical scanners require sunlight shading and protection from wind-blown dust and rain. Scanners can fail to operate in very high ambient temperatures due to sensitive components. Air moisture can also cause interference and laser detection issues

(continued)

Table 2.1 (continued)

Issue	Photogrammetry	Optical laser scanning
Accuracy and completeness	Dependent on the quality and number of images obtained and the software used to produce the model. Undercut areas can cause problems as can deep prints causing shade problems at the bottom of the print. For accurate measurements images have to be carefully scaled	Dependent upon the make and model of the scanner. Difficult to capture undercut or overhanging areas with a vertically mounted scanner; multiple shots may be required and there still may be problems with very deep prints. Scans are scaled accurately as they are captured, provided the scanner is regularly calibrated
Intra- and inter-site variability	The accuracy of a photo-model is specific to one object and the images taken, there is therefore a strong risk of undetected intra- and inter-site variability in accuracy and reliability of the models. The accuracy of every single model needs to be checked via a reference object in every model	Provided a scanner is well-maintained and regularly calibrated by the manufacturer its accuracy should be consistent in intra-site setting and inter-site settings subject to a caveat around changing environmental conditions. The accuracy of scanned images needs only to be checked once at a site, or following best practice daily at most
Edge effects	Taking images close to an excavation wall can be problematic since a full 360° array of images may not be possible	Depends on tripod or frame configuration, but potentially not a problem especially if oblique scans are also used
Risks of failure	Data quality – moderate to high, associated with failure to capture sufficient images of good quality and coverage especially when post-processing is being done on return from the field. Equipment – low since cameras are ubiquitous on field expeditions so multiple options are often available when one camera fails assuming flexible camera mounts and tripod connections. Post-processing – moderate to high, failure of the software to produce adequate models	Data quality – low in terms of failure to capture data since the quality of a model can be instantly verified and checked in the field and scans re-shot if needed. Equipment – moderate to high since scanners are relatively delicate scientific equipment and field failure is usually terminal since few projects have access to multiple scanners. This is low for scanners designed for field use. Post-processing – low focused simply on data quality and enhancement

is required to produce either multiple shots around an image or ones from fixed points depending on the requirements of the software being used. The level of computing requirements also varies but is usually relatively high compared to the average home computer. Shading of ambient light is usually necessary to remove shadow effects and problems may be encountered with deeply impressed prints. Uniform substrate textures, especially under intense sunlight, may also limit the accuracy and reliability of some photo-models. Damage to the site may occur due to tripod/frame legs or by standing/crouching on delicate surface in order to take multiple oblique shots, which also may not always be possible due to excavation walls. The principle risk is that the digital elevation models are post-processed and therefore faults are usually determined once a field scientist has left the field. While in many situations one can return to the field this is not always possible if the subject has been lost to erosion or is exposed at a remote location. The lack of good digital models for the prints at Happisburgh is a case in point (Ashton et al. 2014).

In contrast, optical laser scanners involve a greater capital investment and are more complex to deploy in the field due to power requirements. Most scanners which are designed for engineering or medical purposes have to be protected and mounted within custom built rigs to allow field deployment. Once deployed, a scanner can give fast, accurate and reliable results across a range of surface textures, right up to the edge of an excavation (Fig. 2.9). Data quality and accuracy can be checked in the field and scans re-shot if necessary, minimising risks. Risks of equipment failure are however higher given that scanners are relatively delicate scientific equipment. The authors remember keenly shorting a scanner in northern Kenya, transported at great cost, on day-one of a field expedition when it was plugged into a faulty generator that produced a power spike exploding the scanner and setting light to the associated laptop!

What is clear from the work of Bennett et al. (2013) is that there is currently no perfect solution and field practitioners need to be aware of the rival merits of both optical laser scanning and photogrammetry (Table 2.1). Where the highest standards of accuracy and reliability are required either because of a remote location or because the tracks will only be exposed in an optimal state once, for example upon first excavation, then the use of optical laser scanning supplemented by photogrammetry is perhaps best. Where tracks are less fragile, more accessible and a greater degree of intra-track variability is acceptable then photogrammetry provides a rapid and flexible solution, and is particularly ideal for initial reconnaissance type work. As the sophistication of photogrammetry increases with further software developments and enhanced user interfaces it is likely that long term it will provide a more reliable and efficient field based solution than optical laser scanning, but we are perhaps not there quite yet.

2.5 Data Manipulation

Whether data is obtained via photogrammetry or from an optical laser scanner the principle output is likely to be some form of point cloud consisting of x, y and z coordinates. While it is possible to analyse this in a wide range of commercial three-dimensional software packages there are some simple tools available as freeware which allow one to undertake both basic and sophisticated analyses. At its simplest the first step one may need to take is to scale a point cloud derived from photogrammetry which can be done easily within MeshLab [http://meshlab.sourceforge.net]

which is a freeware package for viewing three-dimensional files and provides options for visualising, surfacing and rectifying the geometry of point clouds. The authors prefer, however, to handle the data within their own freeware Foot Processor [http://footprints.bournemouth.ac.uk/]. It is a piece of bespoke software that allows rapid visual editing of x, y, z data files in order to: (1) rectify tracks to the orthogonal plane for analysis; (2) rotate prints into a consistent longitudinal orientation; (3) mirror left into right prints; (4) invert prints such that high points become lows; (5) crop extraneous material from the margins of a print either via a square, polygon or by contour; and (6) produce contour plots, place landmarks and exports inter-landmark distances and coordinates. There is also a separate tool for viewing multiple footprint files (Foot Viewer) [http://footprints.bournemouth.ac.uk/].

2.6 Basic Measurements: Tracks and Trackways

Whether working in the field, or subsequently on digital elevation models, some form of basic track measurement will be required. Figure 2.11 defines the basic measurements associated with gait (Levine et al. 2012), including stride and step length and these are reviewed in detail by Wilkinson et al. (1995) who recommends the use of a line of progression based on ipsilateral measurements. In terms of basic track dimensions there is a lack of historic consistency and different practices exist between clinical, forensic and anthropological disciplines. Robbins (1985) recognised this and tried to bring consistency to these measurement schemes. Robbins defined a Designated Longitudinal Axis (DLA) for a track that stretches from an arbitrary point between the first and second toes and the most proximal point of the heel (Pternion). She defined foot length in various ways, recognising a number of landmarks around the margins of the foot based on skeletal protuberance and developed a complex classification of foot shape (Fig. 2.12a). While Robbins' aim was laudable, to bring consistency where none existed, the resulting system was simply to complex and the skeletal landmarks unrecognisable in many tracks. As a consequence it has not been adopted by all practitioners. Gunn (1991) reviewed the literature and favoured the use of five foot lengths (Gunn lines) from the Pternion to the end of each toe, which was one of the approaches advocated by Robbins (1985) and now underpins much of the forensic literature (see Sects. 6.2 and 7.3). Some workers adopt a more formal definition of the central axis and the Pternion (Reel et al. 2010, 2012; Fig. 2.12b). Others favour the use of dimensions based on defining the centroid of such things as the heel or toe pads rather than the edge or some combination of both (Kennedy et al. 2005). The advantage of the centroid in tracks studies is that it takes the measurements away from the influence of track walls which in deeper tracks may reflect the sides of the foot rather than the plantar surface and sub-vertical walls may pose a challenge for landmark placement (Appendix, Track 1).

In track studies quantities such as foot length and width are often poorly defined and are usually based on maximum dimensions (e.g., Roberts et al. 1996; Schmincke et al. 2010). The authors favour a modified version of the Robbins's (1985) scheme

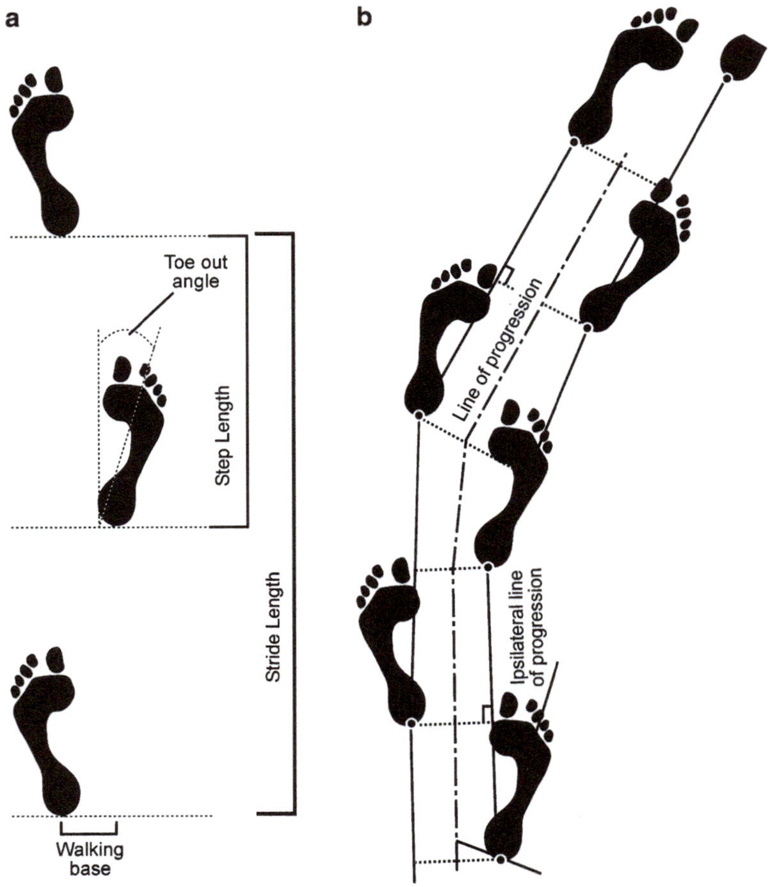

Fig. 2.11 Measures of gait including definitions of step length, stride length and toe-out angle after Levine et al. (2012) and Wilkinson et al. (1995)

in which the longitudinal axis is defined as that between the Pternion and the maximum extent of the second toe and gives a longitudinal axis that can be defined quickly and consistently in the field unlike that used by Reel et al. (2010, 2012). Foot length is taken along this axis and measures of heel and ball width are maximum values approximately perpendicular to the longitudinal axis (Fig. 2.12d). While it would be ideal for all researchers to use a common set of measures, this is in practice unlikely and in light of this it is important that authors clearly define their chosen landmarks and that in making comparisons readers acknowledge that 'apples and pears' may be the order of the day. Measures of foot length do correlate well with one another on true tracks and any variance caused by different length definitions for example is likely to be small compared to that associated with intra-trackway variation. Using Harry's Trackway (Walvis Bay Namibia; see Sects. 1.4 and 3.2.3)

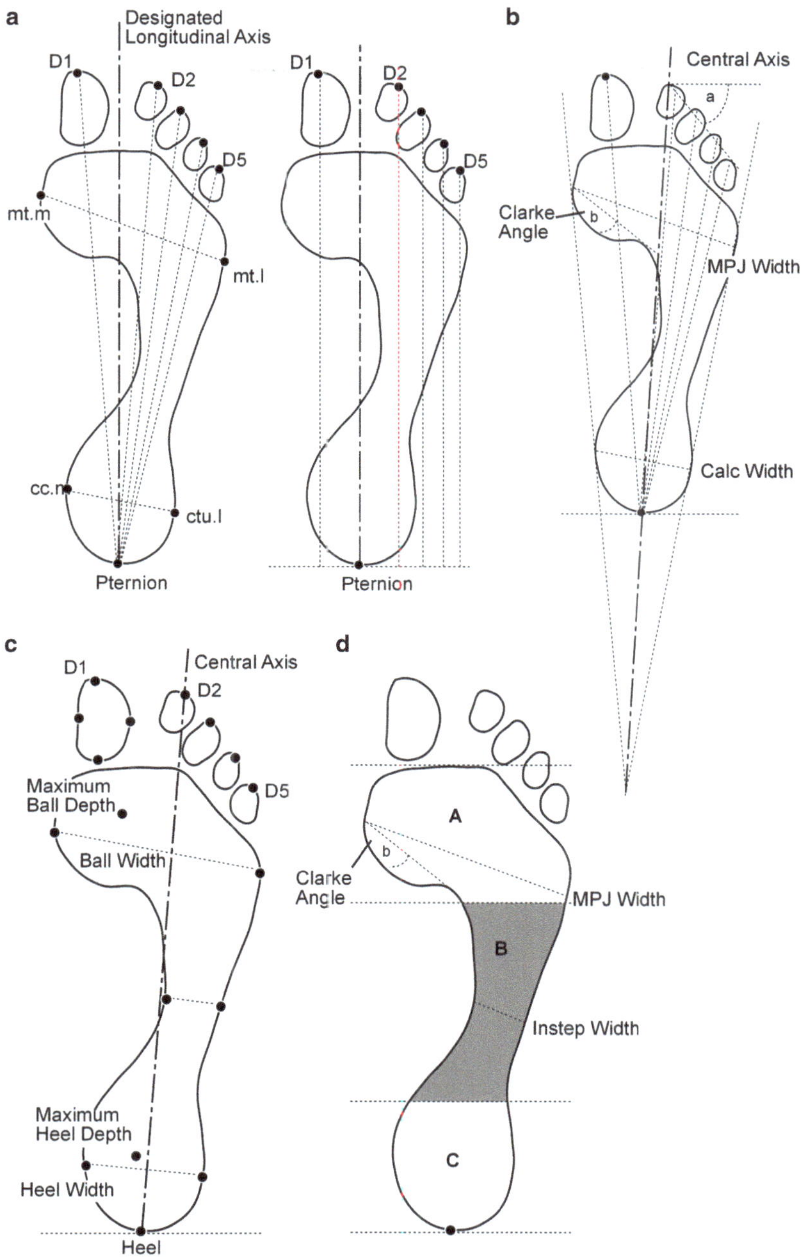

Fig. 2.12 Various measurement systems for human tracks. (**a**) Measurement system developed by Robbins (1985). (**b**) Measurement system based on Gunn (1991) and adopted by Reel et al. (2010, 2012). (**c**) Landmarks and measurement scheme favoured by the authors. (**d**) Various measures of longitudinal arch development including the Clarke Angle

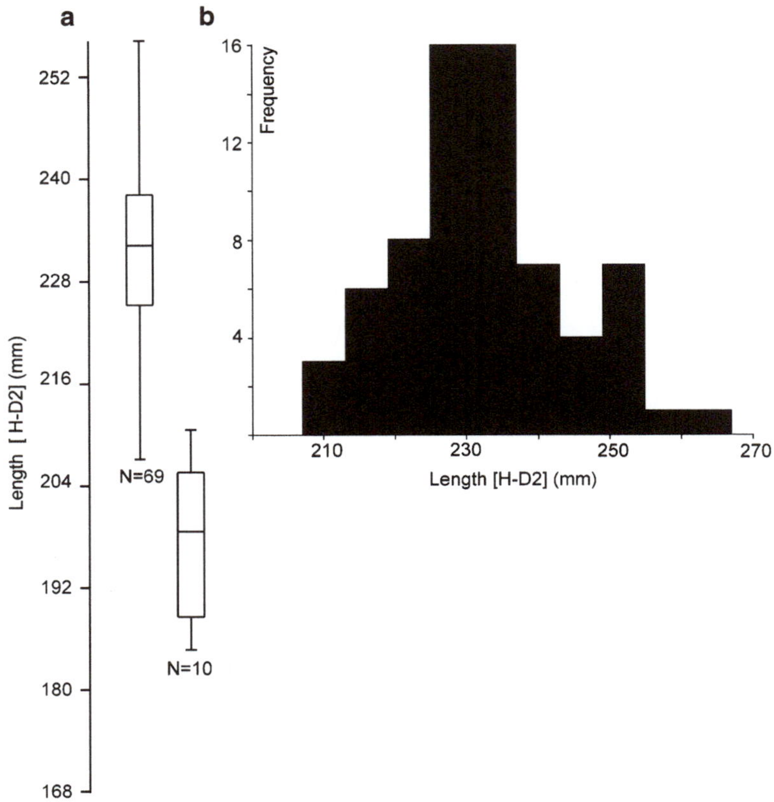

Fig. 2.13 Variability in foot length within Harry's and Harriet's trackways. (**a**) Mean and ranges for both trackways. (**b**) Frequency histogram of track lengths within Harry's Trackway

the variation in track length is quite marked as shown in Fig. 2.13. This variation is a function of variations in substrate, proximal foot slippage and in some cases distal drag marks as the toes lift.

Within the clinical and forensic literature there are a number of additional foot measurements that are used to quantify in particular the degree of development of the longitudinal medial arch for example. The key assumption here is that foot shape correlates with the medial longitudinal arch of the foot in some way (e.g., Cavanagh and Rodgers 1987; Gilmour and Burns 2001). This idea lies behind Robbins's (1985) shape classification and Stavlas et al. (2005) proposed a series of foot types based on a numerical foundation (Fig. 2.14). Schwartz et al. (1928) advocate a measure known as the footprint angle or Clarke Angle after the additional work of Clarke (1933); the bigger this angle the better developed the arch (Fig. 2.12e). Cavanagh and Rodgers (1987) introduced something known as the arch-index, essentially the ratio of the mid-foot area to the whole area (Fig. 2.12e), while the Chippaux-Smirak index based on a

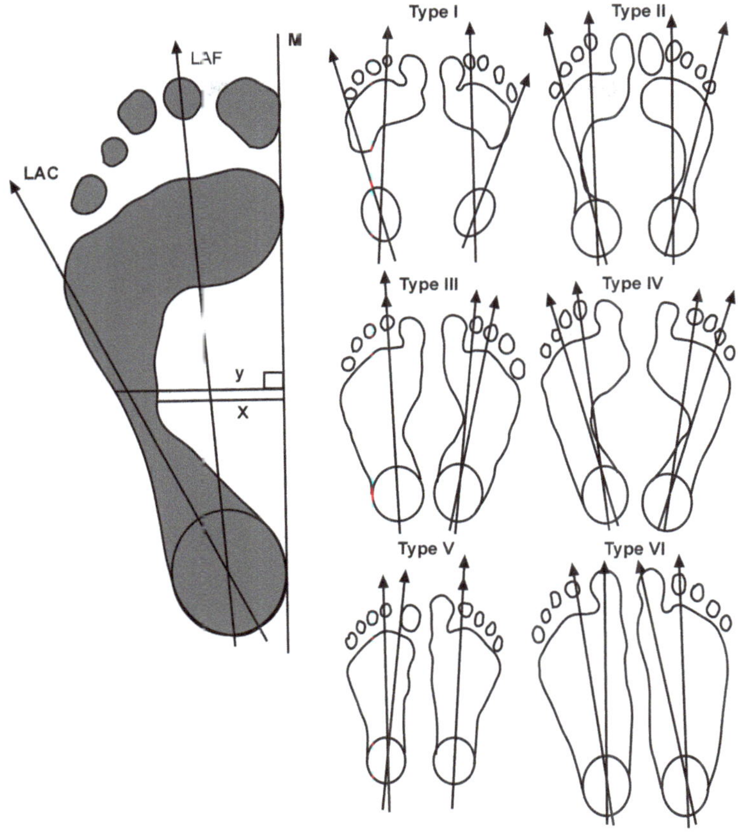

Fig. 2.14 Scheme for recording arch development in feet developed for two-dimensional tracks but potentially applicable to three-dimensional tracks following and modified from Stavlas et al. (2005). A typical footprint and the six footprint types are shown. The longitudinal axis of the foot (*LAF*) is the line from the centre of hind-foot imprint to the second toe. The longitudinal axis of calcaneus imprint (*LAC*) is the line that bisects the imprint of the hind-foot defined by an oval. In every footprint, a line (*M*) is drawn along the medial border of the foot. A perpendicular line (*y*) is then drawn, from line (*M*) to the lateral outline of the isthmus, corresponding to the width of mid-foot. An additional line (*x*), which is parallel to (*y*), is drawn in the mid-foot, corresponding to the width of the arch. The difference between y and x gives the width of the isthmus and along with its relationship to the LAF and LAC and allows six foot types to be defined

ratio of the ball width to mid-foot width is another similar index (Stavlas et al. 2005). In the context of tracks, arch development is often obscured by the movement of sediment in a proximal direction within a footprint (Brown 1999; Appendix, Track 4) and is not always as evident as found in clinical/forensic studies where two-dimensional footprints based on pressure sensitive paper or force plates are common. It is not surprising therefore that many of the measures developed within the clinical and forensic literature have yet to be applied to the study of the human tracks in the geological

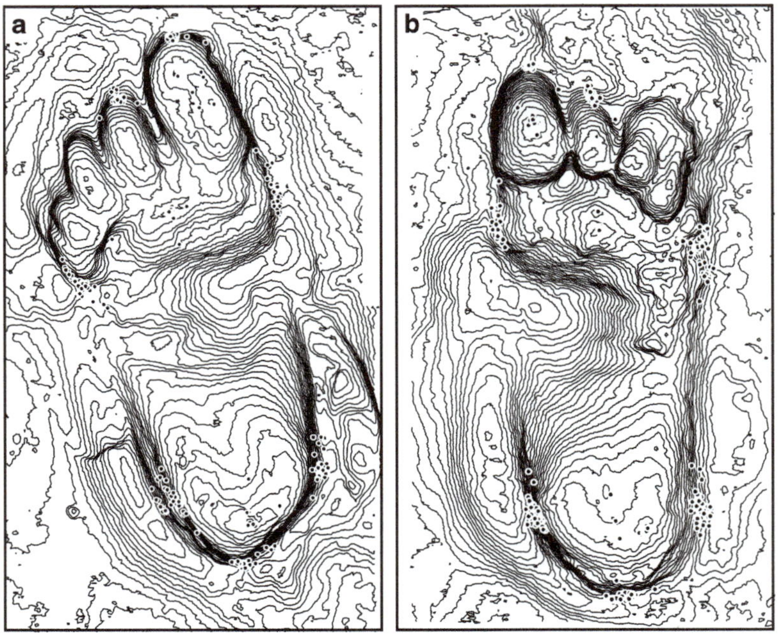

Fig. 2.15 Landmark experiment based on two tracks from Harry's Trackway. Two contour maps of footprints were given to 50 Anthropology Masters students at Bournemouth University in 2011. The students were asked to place ten landmarks; one to mark the pternion, the extent of the first and second toes, the deepest point of the first toe, the maximum width and deepest part of the heel, the maximum ball wall width and the deepest point in the ball area. As you can see there is wide range of landmark placements especially around the medial and lateral margins of the tracks

record. It is worth emphasising that all linear measures of a human track require the placement of a landmark, whether a point on a digital track model, or mentally when taking field measurements. Landmarks are subject to inter-operator errors as illustrated in Fig. 2.15 which shows the placement of a series of landmarks on two tracks from Harry's Trackway.

2.7 Advanced Measurements: Tracks and Trackways

Hypothesis testing and comparison of the anatomy of different populations of human tracks has until recently been limited to largely qualitative approaches. Berge et al. (2006) pioneered the application of geometric morphometrics to the analysis of human tracks (Meldrum and Chapman 2007) an approach adopted and developed further by Bennett et al. (2009) in the analysis of the Ileret footprints. Geometric morphometric tools attempt to preserve the geometry of a structure during analysis such that the statistical representation and comparisons of shape is possible independent, at least in theory, of size (Richtsmeier et al. 2002; Gómez-Robles et al. 2008; Mitteroecker and Gunz 2009; Friess 2010; Webster and Sheets 2010; Zelditch et al. 2012). The simplest application and that used by Berge et al. (2006) is based on defining homology-based landmarks (Slice 2007; Polly 2008); that is those that relate to biologically or anatomically homologous structures and crucially can be recognised consistently by observers. The resultant coordinates (two- or three-dimensional) are used to explore inter-landmark distances and can be transformed using such tools as a Generalised Procrustes Analysis to effectively scale, translate and transform the landmark configurations of individual subjects into a common coordinate space such that anatomical properties of shape can be explored in subsequent multivariate analyses. This is shown schematically in Fig. 2.16a and allows track populations to be compared statistically and patterns of intra- or inter-population landmark variation to be mapped. These approaches have been widely applied within palaeontology and palaeoanthropology. The freeware PAST (http://folk.uio.no/ohammer/past) linked to a textbook by Hammer and Harper (2006) places these tools in the hands of most palaeontologists and has been widely used in track studies by the authors. Bennett et al. (2009) in their initial paper on the Ileret footprints used a landmark based approach. Figure 2.17 shows a landmark based comparison showing the shape difference between various populations in this analysis and the data was used in a discriminant analysis in which two end-members were provided by modern tracks (*Homo sapiens*) and those at Laetoli (*Australopithecus afarensis*). The Ileret tracks were classified in this model and found to be indistinguishable from the modern tracks, while the recognition of homologous landmarks across species is not without its challenges the analysis was the first quantitative inter-species assessment of human tracks.

We can illustrate the approach further by using Harry's Trackway (Walvis Bay Namibia; see Sects. 1.4 and 3.2). In this illustration a series of landmarks have been placed on the individual tracks and subject to a Generalised Procrustes Analysis. Figure 2.18 shows a deformation grid of the first Principle Component from a shape-based Principle Components Analysis, which demonstrates that around 26 % of the variance in the landmark location in this analysis can be accounted for in the position of the deepest point of the heel and ball areas of the track and variation in the degree of posterior placement of the fifth toe pad. The second component depicts variation in the mid-foot, essentially the degree of development of the medial longitudinal arch. Figure 2.18c shows the variation relative difference in mean shape as a deformation grid between Harry's and Harriet's trackways.

The problem with such approaches, in the context of studying tracks, is that landmarks work well where anatomically significant points can be recognised and defined consistently, but this is not always the case where areas or regions of interest may be more relevant. The landmarks placed in Fig. 2.18 for example are for the most part perimeter based, yet the interesting variation occurs inside that perimeter and consistent landmark placement is much harder in these areas. In these cases the use of homology-free landmark systems may be more appropriate in which their location is determined by some form of mathematical algorithm or principle (Gunz et al. 2005; Polly 2008; Gunz and Mitteroecker 2013), although Klingenberg (2008)

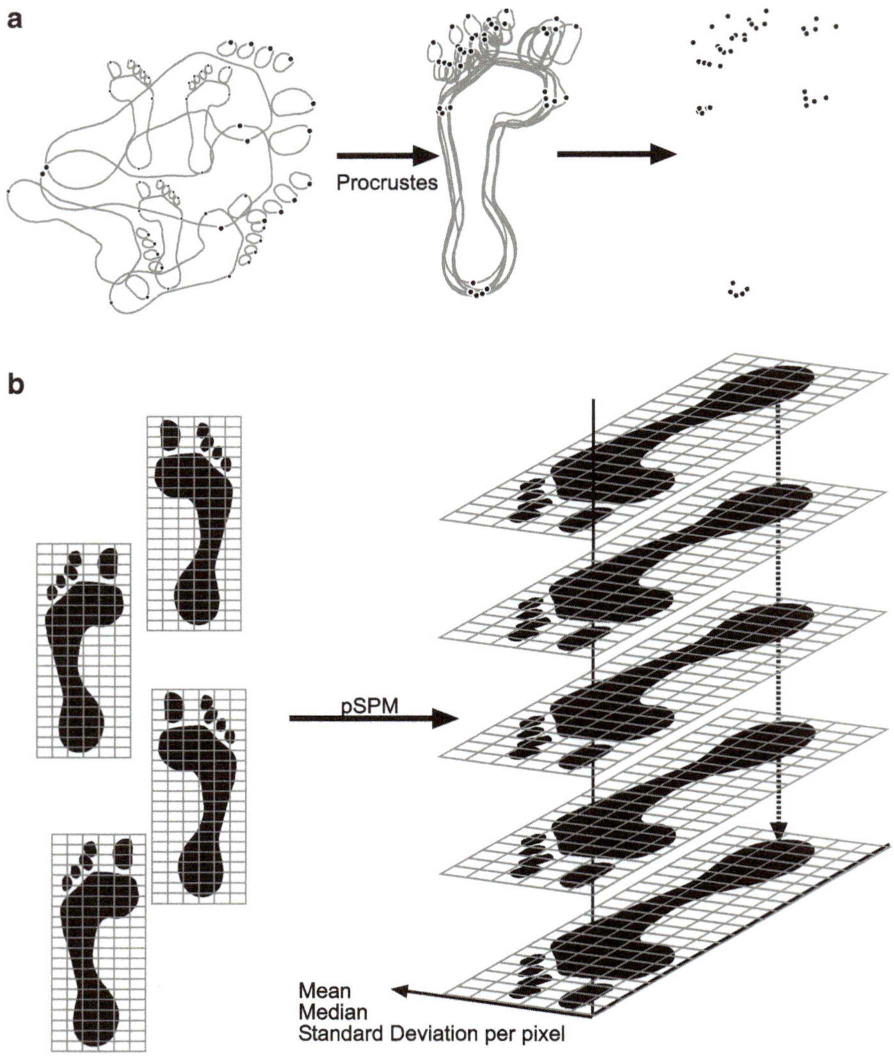

Fig. 2.16 Different approaches shown schematically for the advanced analysis of human tracks. (**a**) Use of landmarks and Generalised Procrustes Analysis. (**b**) The conceptual foundation of Pedobarographic Statistical Parametric Mapping (pSPM). See text for further discussion

argues that such methods are not, free of some kind of anatomical assumptions. Such approaches may offer an alternative approach for smooth surface areas, such as the base of a track, which may be 'landmark free' and were differences depend on subtle variations in surface curvature (e.g., Perez et al. 2006; Slice 2007). Here the use of sliding-landmarks or semi-landmarks may be appropriate. These are points assigned at regular intervals along a line or across a surface and the standard Procrustes Analysis is extended so that in addition to translating, scaling and rotat-

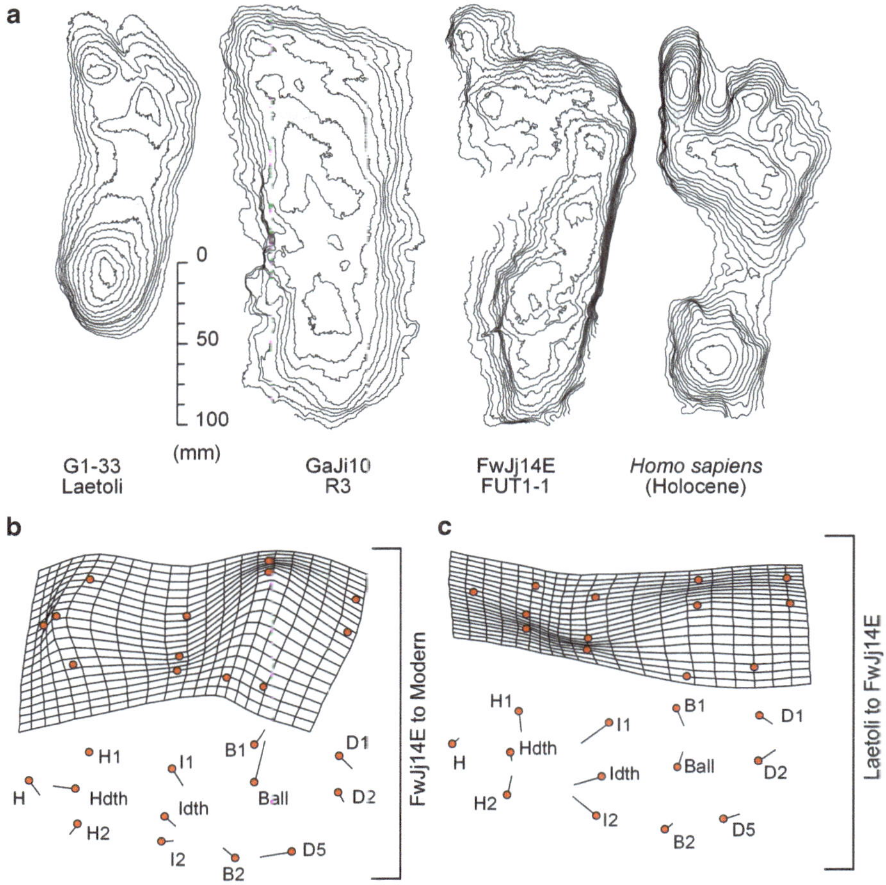

Fig. 2.17 Landmark based analysis of the Ileret footprints from northern Kenya following Bennett et al. (2009)

ing the landmarks optimally, they are also allowed to slide along the curve or across the surface until they match in an ideal fashion the positions of corresponding points on a reference form (Adams et al. 2004; Perez et al. 2006). These approaches have yet to be applied to human tracks. It is also worth noting that Sforza et al. (2000) used Fourier analysis to explore the bilateral asymmetry of two-dimensional tracks and such approaches may also yield valuable results in the future (Hammer and Harper 2006). The other challenge with landmark based approaches is the incorporation of specimens where there are missing landmarks, for example in the case of partial track, which is something that has been subject to considerable research in the field of geometric morphometrics in recent years (Adams et al. 2004; Slice 2007). This whole field is full of potential applications to the study of human tracks and deserves further exploration in the future.

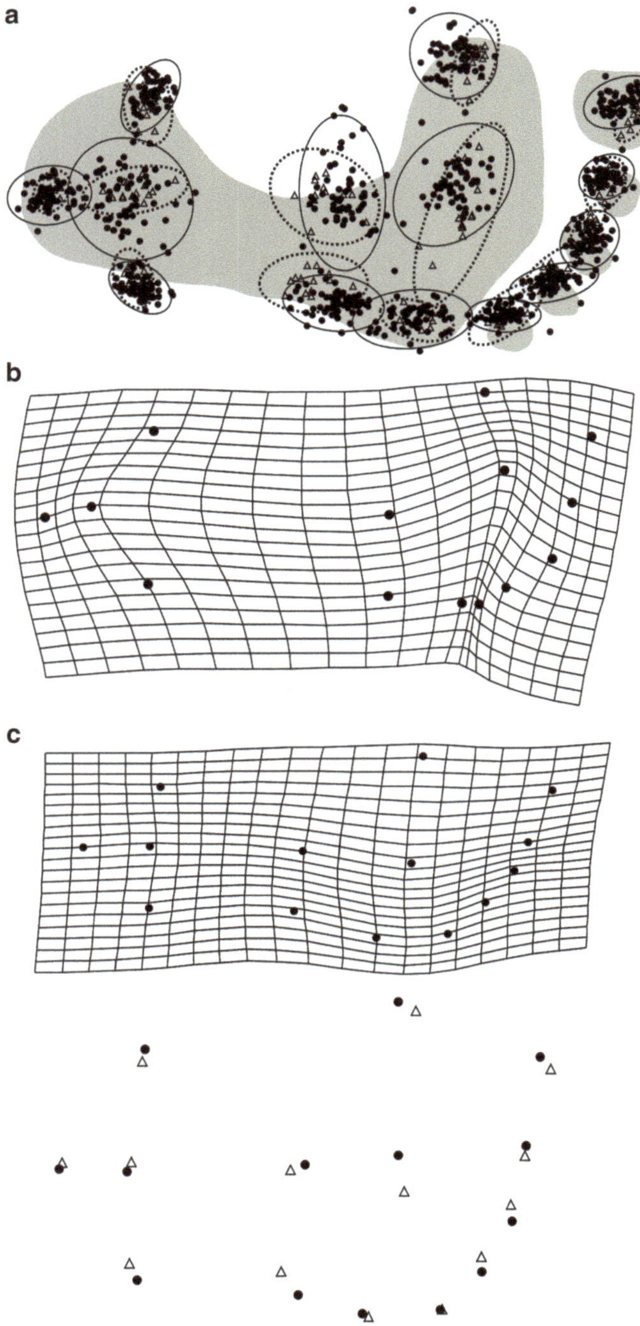

Fig. 2.18 Application of geometric morphometrics to Harry's and Harriet's trackways. (**a**) The distribution of landmarks for Harry's (*dots*) and Harriet's (*triangles*) trackways with 95 % confidence limit ellipses shown. The landmarks have been subject to a Generalised Procrustes Analysis in PAST. (**b**) First component (26 % of the variance) of a Principal Components Analysis of the landmark distribution showing that most of the variation is associated with: the location of the deepest point in the heel; the location of the area of maximum depth in the ball area; and the placement of the firth toe pad. (**c**) A simple thin-plate spline comparison of the mean landmarks for Harry's and Harriet's trackways shown as a deformation grid and simply as landmarks

A different approach based on a 'whole-foot' analysis, and therefore dispensing with landmarks, has been developed for plantar pressure records and recently applied to human tracks substituting depth for pressure. Pedobarographic Statistical Parametric Mapping (pSPM) computes measures of central tendency for multiple pressure records obtained from a pressure treadmill (Pataky and Goulermas 2008; Pataky et al. 2008a, b). It is based on the idea that multiple footprints have similar geometry allowing them to be registered (Maintz and Viergever 1998), or spatially transformed to match a template foot or record (Pataky et al. 2008b). If registered correctly, each pixel should correspond to the same anatomical location in all co-registered images (Pataky et al. 2008a, b; Fig. 2.16b). This allows measures of central tendency to be computed for each pixel allowing mean or median records to be obtained for a given population. Within the pSPM software developed by Pataky and his colleagues registration is either achieved automatically through a sequence of trial registrations or can be done manually where the pressure records are more variable. It is possible to extend this technique in order to undertake pixel-wise two-sample t tests (Friston et al. 2007; Crompton et al. 2012) and thereby compare two means statistically. While the approach was developed for pressure records it has been used for tracks most notably in comparing the Laetoli tracks with modern track populations (Crompton et al. 2012). Although mathematically complex and not currently available openly to the research community pSPM, or for that matter any similar whole-foot approaches, offers a number of benefits for footprint studies by allowing the objective testing of hypotheses using the whole of a foot and frees researchers from the potential subjectivity associated with the selection and placement of landmarks (Crompton et al. 2012). Figure 2.19 shows a mean track along with measures of central tendency for both Harry's and Harriet's trackways.

The approach is not however without significant challenges. Automatic registration within pSPM requires a level of homology between prints, such that the difference can be minimised through registration, leaving only those of anatomical significance. This is easy for pressure records since they are topologically consistent, however human tracks aren't and the margins of prints often vary quite markedly within a trackway. In order to remove this unwanted noise it is necessary within pSPM to discard everything but the plantar surface by applying some form of vertical threshold to delete the unwanted data. While this focuses attention on the plantar surface it is no longer a 'whole-foot' approach and geologically the peripheral areas of a print may be very important to interpreting the interaction of a foot with the substrate (see Sect. 5.2). The alternative approach is to use a manual registration within pSPM which is possible for small numbers of prints especially in making comparison between potentially different hominin species (Crompton et al. 2012), but is not possible in the analysis of large data sets and is no longer truly objective. Notwithstanding these issues the 'whole-foot' approach used by Crompton et al. (2012) has huge potential and provides a very clear alternative to landmark based approaches. It also sets the challenge to the ichnological community to explore other ways of co-registering a series of tracks allowing measures of central tendency to be described rather than individual tracks which are subject to significant intra-trackway variability.

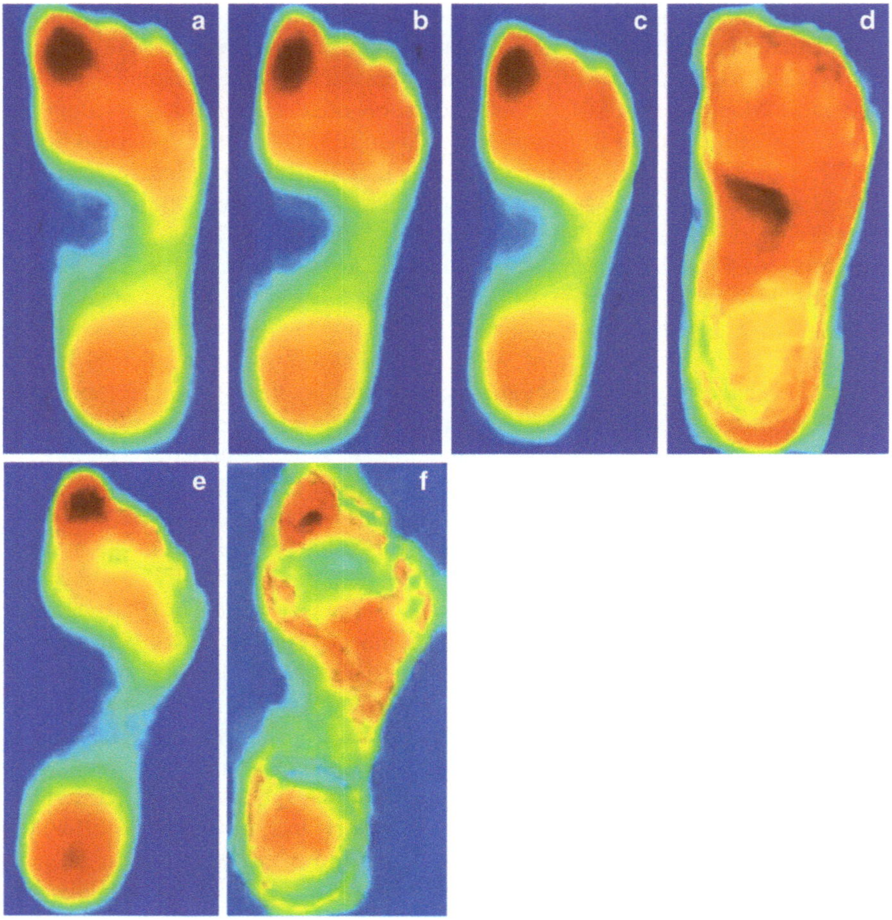

Fig. 2.19 Mean and standard deviations tracks for Harry's and Harriet's trackways produced using Pedobarographic Statistical Parametric Mapping (pSPM). (**a**) Mean right foot Harry's Trackway. (**b**) Mean Left foot – shown as a right foot – for Harry's Trackway. (**c**) Mean of both right and left feet of Harry's Trackway based on 54 individual tracks from the trackway. (**d**) Standard deviation showing the variation within the trackway. (**e, f**) Mean and standard deviation for Harriet's Trackway based on a total of ten tracks

2.8 Summary

The methods that are available with which to document human tracks have become increasingly sophisticated in recent years. The collection of three-dimensional data whether by optical laser scanning or photogrammetry should now be considered standard for any investigation and is the key to sophisticated analysis or hypothesis testing. It is also something that is now within the grasp of every researcher. However fundamental to the process is geo-prospection since one must first find the tracks and it is the contention of the authors' that human and animal tracks are perhaps more common in the geological record than often assumed and that in many cases they are simply not recognised in vertical cross-section. Certainly in environments such as those the Turkana Basin in northern Kenya this may well be true and there are lots more tracks to be found here in the future. The prominence of tracks, such as those of Laetoli in northern Tanzania, found within volcanic ash tends in some respects to focus attention away from sites where most human tracks have been found; those associated with fine-grained sediment as part of fluvial and lacustrine depositional settings. Dating of tracks is also an important task and one not without potential controversy. We have seen how there is also a diversity of measurement schemes applicable to human tracks and a need for the community and its various component disciplines to find great methodological consistency. At the same time there is huge potential to apply the principles and techniques of geometric morphometrics to human tracks and for the community to develop different types of 'whole-foot' analytical procedures. The prospect is huge and in the next chapter we explore the diverse range of existing track sites in order to illustrate the range of potential environments that exist and the challenges that one may face in studying them.

References

Adams DC, Rohlf J, Slice DE (2004) Geometric morphometrics: ten years of progress following the 'revolution'. Ital J Zool 71:5–16

Allen JRL (1989) Short paper: Fossil vertebrate tracks and indenter mechanics. J Geol Soc 146(4):600–602

Allen JRL (1997) Subfossil mammalian tracks (Flandrian) in the Severn Estuary, S.W. Britain: mechanics of formation, preservation and distribution. Philos Trans R Soc Lond B B352:481–518

Ashton N, Lewis SG, De Groote I et al (2014) Hominin footprints from early Pleistocene deposits at Happisburgh, UK. PLoS One 9(2):e88329. doi:10.1371/journal.pone.0088329

Bennett MR, Harris JWK, Richmond BG et al (2009) Early hominin foot morphology based on 1.5 million year old footprints from Ileret, Kenya. Science 323:1197–1201

Bennett MR, Gonzalez S, Huddart D et al (2010) Probable Neolithic footprints preserved in inter-tidal peat at Kenfig, South Wales (UK). Proc Geol Assoc 121:66–76

Bennett MR, Falkingham P, Morse SA et al (2013) Preserving the impossible: conservation of soft-sediment hominin footprint sites and strategies for three-dimensional digital data capture. PLoS One 8(4):e60755. doi:10.1371/journal.pone.0060755

Berge C, Penin X, Pellé É (2006) New interpretation of Laetoli footprints using an experimental approach and Procrustes analysis: preliminary results. Comptes Rendus Palevol 5:561–569

Breithaupt BH, Matthews NA, Noble TA (2004) An integrated approach to three-dimensional data collection at dinosaur tracksites in the Rocky Mountain West. Ichnos 11(1–2):11–26

Brown T (1999) The science and art of tracking – nature's path to spiritual discovery. Berkley Books, New York

Cavanagh PR, Rodgers MM (1987) Pressure distribution under symptom-free feet during barefoot standing. Foot Ankle Int 7(5):262–278

Cho DL, Park KH, Jin JH et al (2005) Age constraints on human footmarks in Hamori Formation Jeju Island, Korea. J Petrol Soc Korea 14:149–156

Clarke HH (1933) An objective method of measuring the height of the longitudinal arch in foot examinations. Res Q Am Phys Educ Assoc 4(3):99–107

Crompton RH, Pataky TC, Savage R et al (2012) Human-like external function of the foot, and fully upright gait, confirmed in the 3.66 million year old Laetoli hominin footprints by topographic statistics, experimental footprint-formation and computer simulation. J R Soc Interface 9:707–719

Day MH, Wickens EH (1980) Laetoli pliocene hominid footprints and bipedalism. Nature 286:385–387

de Gilbert JM, Sáez A (2009) Paleohydrological significance of trace fossil distribution in Oligocene fluvial-fan-to-lacustrine systems of the Ebro Basin, Spain. Palaeogeogr Palaeoclimatol Palaeoecol 272(3):162–175

Falkingham PL (2012) Acquisition of high resolution three-dimensional models using free, open-source, photogrammetric software. Palaeontol Electronica 15:1T:15p

Falkingham PL, Bates KT, Farlow JO (2014) Historical photogrammetry: Bird's Paluxy River Dinosaur chase sequence digitally reconstructed as it was prior to excavation 70 years ago. PLoS One 9(4):e93247. doi:10.1371/journal.pone.0093247

Feinberg JM, Renne PR, Arroyo-Cabrales J, Waters MR, Ochoa-Castillo P, Perez-Campa M (2009) Age constraints on alleged "footprints" preserved in the Xalnene Tuff near Puebla, Mexico. Geology 37(3):267–270

Friess M (2010) Calvarial shape variation among Middle Pleistocene hominins: an application of surface scanning in palaeoanthropology. Comptes Rendus Palevol 9(6):435–443

Friston KJ, Ashburner JT, Kiebel SJ et al (2007) Statistical parametric mapping: the analysis of functional brain images. Elsevier/Academic, Amsterdam

Gilmour JC, Burns Y (2001) The measurement of the medial longitudinal arch in children. Foot Ankle Int 22(6):493–498

Gómez-Robles A, Martinón-Torres M, Bermúdez de Castro JM et al (2008) Geometric morphometric analysis of the crown morphology of the lower first premolar of hominins, with special attention to Pleistocene *Homo*. J Hum Evol 55(4):627–638

González S, Huddart D, Bennett MR et al (2006) Human footprints in Central Mexico older than 40,000 years. Quat Sci Rev 25(3):201–222

Gunn N (1991) Old and new methods of evaluating footprint impressions by a forensic podiatrist. Br J Poediatry Med Surg 3:8–11

Gunz P, Mitteroecker P (2013) Semilandmarks: a method for quantifying curves and surfaces. Hystrix Ital J Mammal 24(1):103–109

Gunz P, Mitteroecker P, Bookstein FL (2005) Semilandmarks in three dimensions. In: Slice DE (ed) Modern morphometrics in physical anthropology. Springer, Boston

Hammer Ø, Harper D (2006) Paleontological data analysis. Blackwells, Oxford

Kennedy RB, Chen S, Pressman IS et al (2005) A large-scale statistical analysis of barefoot impressions. J Forensic Sci 50(5):1071–1080

Kim CB, Kim JY, KIM KS et al (2010) New age constraints for hominid footprints found on Jeju Island, South Korea. J Arch Sci 37:3338–3343

Kinahan J (1996) Human and domestic animal tracks in an archaeological lagoon deposit on the coast of Namibia. South Afr Archaeol Bull 51:94–98

Klingenberg CP (2008) Novelty and "homology-free" morphometrics: what's in a name? Evol Biol 35:186–190

Krapovickas V, Ciccioli PL, Mángano MG et al (2009) Paleobiology and paleoecology of an arid–semiarid Miocene South American ichnofauna in anastomosed fluvial deposits. Palaeogeog Palaeoclimatol Palaeoecol 284(3):129–152

Laporte LF, Behrensmeyer AK (1980) Tracks and substrate reworking by terrestrial vertebrates in Quaternary sediments of Kenya. J Sediment Res 50:1337–1346

Laury RL (1980) Paleoenvironment of a late Quaternary mammoth-bearing sinkhole deposit, Hot Springs, South Dakota. Geol Soc Am Bull 91(8):465–475

Lea PD (1996) Vertebrate tracks in Pleistocene eolian sand-sheet deposits of Alaska. Quat Res 45(2):226–240

Leakey MD (1978) Pliocene footprints at Laetoli, northern Tanzania. Antiquity 52:133
Leakey MD, Harris JM (1987) Laetoli: a Pliocene site in northern Tanzania. Clarendon, Oxford
Levine D, Richards J, Whittle MW (2012) Whittle's gait analysis, 5th edn. Elsevier Health Sciences, London
Lockley M, Roberts G, Kim JY (2008) In the footprints of our ancestors: an overview of the hominid track record. Ichnos 15:106–125
Loope DB (1986) Recognizing and utilizing vertebrate tracks in cross section: Cenozoic hoofprints from Nebraska. Palaios 1(2):141–151
Maintz JBA, Viergever MA (1998) A survey of medical image registration. Med Image Anal 2:1–37. doi:10.1016/S1361-8415(01)80026-8
Mastrolorenzo G, Petrone P, Pappalardo L et al (2006) The Avellino 3780-yr-BP catastrophe as a worst-case scenario for a future eruption at Vesuvius. Proc Natl Acad Sci 103:4366–4370. doi:10.1073/pnas.0508697103
McDougal IAN, Brown FH (2006) Precise $^{40}Ar/^{39}Ar$ geochronology for the upper Koobi Fora Formation, Turkana Basin, northern Kenya. J Geol Soc 163:205–220
Melchor RN, Bedatou E, de Valais S et al (2006) Lithofacies distribution of invertebrate and vertebrate trace-fossil assemblages in an early Mesozoic ephemeral fluvio-lacustrine system from Argentina: implications for the Scoyenia ichnofacies. Palaeogeogr Palaeoclimatol Palaeoecol 239:253–285
Meldrum DJ, Chapman RE (2007) Morphometrics of the outline shape of hominid footprints. Am J Phys Anthropol Suppl 42:170. doi:10.1002/ajpa
Meldrum DJ, Lockley MG, Lucas SG et al (2011) Ichnotaxonomy of the Laetoli trackways: the earliest hominin footprints. J Afr Earth Sci 60(1):1–12
Mitteroecker P, Gunz P (2009) Advances in geometric morphometrics. Evol Biol 36:235–247
Morse SA, Bennett MR, Gonzalez S et al (2010) Techniques for verifying human footprints: reappraisal of pre-Clovis footprints in Central Mexico. Quat Sci Rev 29(19):2571–2578
Morse SA, Bennett MR, Liutkus-Pierce C et al (2013) Holocene footprints in Namibia: the influence of substrate on footprint variability. Am J Phys Anthropol 151:265–279
Pataky TC, Goulermas JY (2008) Pedobarographic statistical parametric mapping (pSPM): A pixel-level approach to foot pressure image analysis. J Biomech 41:2136–2143
Pataky TC, Caravaggi P, Savage R et al (2008a) New insights into the plantar pressure correlates of walking speed using pedobarographic statistical parametric mapping (pSPM). J Biomech 41:1987–1994
Pataky TC, Goulermas JH, Crompton RH (2008b) A comparison of seven methods of within-subjects rigid-body pedobarographic image registration. J Biomech 41:3085–3089
Perez SI, Bernal V, Gonzalez PN (2006) Differences between sliding semi-landmark methods in geometric morphometrics, with an application to human craniofacial and dental variation. J Anat 208:769–784
Polly PD (2008) Developmental dynamics and G-Matrices: can morphometric spaces be used to model phenotypic evolution? Evol Biol 35:83–96
Raichlen DA, Gordon AD, Harcourt-Smith WE et al (2010) Laetoli footprints preserve earliest direct evidence of human-like bipedal biomechanics. PLoS One 5(3):e9769
Reel S, Rouse S, Vernon W et al (2010) Reliability of a two-dimensional footprint measurement approach. Sci Justice 50:113–118
Reel S, Rouse S, Vernon W et al (2012) Estimation of stature from static and dynamic footprints. Forensic Sci Int 219:283-e1
Richtsmeier JT, Burke Deleon V, Lele SR (2002) The promise of geometric morphometrics. Am J Phys Anthropol 119(S35):63–91
Robbins LM (1985) Footprints: collection, analysis, and interpretation. CC Thomas, Springfield
Roberts G, Gonzalez S, Huddart D (1996) Inter-tidal Holocene footprints and their archaeological significance. Antiquity 70:647–651
Schmincke H-U, Kutterolf S, Perez W et al (2009) Walking through volcanic mud: the 2,100 year old Acahualinca footprints (Nicaragua). Bull Volcano 71:479–493

Schmincke HU, Rausch J, Kutterolf S et al (2010) Walking through volcanic mud: the 2,100 year-old Acahualinca footprints (Nicaragua) II: the Acahualinca people, environmental conditions and motivation. Int J Earth Sci 99(1):279–292

Schwartz L, Britten RH, Thompson LR (1928) Studies in physical development and posture. US Publ Health Bull No. 179. US Government Printing Office, Washington, DC

Scrivner PJ, Bottjer DJ (1986) Neogene avian and mammalian tracks from Death Valley National Monument, California: their context, classification and preservation. Palaeogeogr Palaeoclimatol Palaeoecol 57:285–331

Sforza C, Fragnito N, Serrao G, Ferrario VF (2000) Harmonic analysis of footprint symmetry in healthy adolescents. Ann Anat 182(3):285–291

Slice DE (2007) Geometric morphometrics. Annu Rev Anthropol 36:261–281

Sohn YK, Cronin SJ, Brenna M et al (2012) Ilchulbong tuff cone, Jeju Island, Korea, revisited: a compound monogenetic volcano involving multiple magma pulses, shifting vents and discrete eruptive phases. Geol Soc Am Bull 124:259–274

Sonh YK, Park JB, Khim BK et al (2002) Stratigraphy petrochemistry and Quaternary depositional record of the Songaksan tuff ring, Jeju Island. Korea J Volcanol Geother Res 119:1–20

Stavlas P, Grivas TB, Michas C et al (2005) The evolution of foot morphology in children between 6 and 17 years of age: a cross-sectional study based on footprints in a Mediterranean population. J Foot Ankle Surg 44:424–428

Tuttle RH (1984) Bear facts and Laetoli impressions. Am J Phys Anthropol 63(2):230

Tuttle RH (1987) Kinesiological inferences and evolutionary implications from Laetoli bipedal trails G-1. G-2t3 and A. In: Leakey MD, Harris JM (eds) Laetoli: a Pliocene site in northern Tanzania. Clarendon Press, Oxford, pp 503–523

Tuttle RH (2008) Footprint clues in hominid evolution and forensics: Lessons and limitations. Ichnos 15(3–4):158–165

Tuttle RH, Webb D, Weidl E et al (1990) Further progress on the Laetoli trails. J Archaeol Sci 17:347–362

Van der Lingen GJ, Andrews PB (1969) Hoof-print structures in beach sand. J Sediment Res 39(1):350–357

Webster MARK, Sheets HD (2010) A practical introduction to landmark-based geometric morphometrics. Quant Methods Paleobiol Paleontol Soc Papers 16:163–188

Wilkinson MJ, Menz HB, Raspovic A (1995) The measurement of gait parameters from footprints. Foot 5:84–90

Willey P, Stolen J, Crothers G et al (2005) Preservation of prehistoric footprints in Jaguar Cave, Tennessee. J Cave Karst Stud 67(1):61–68

Zelditch ML, Swiderski DL, Sheets HD (2012) Geometric morphometrics for biologists: a primer. Academic, New York

Chapter 3
World Review of Human Track Sites

Abstract Human tracks have now been recorded at a number of sites across the globe. Lockley et al. (Ichnos 15:106–125, 2008) provides a definitive review of many of these sites and our aim here is to focus on a few important examples which are either in the authors' judgement particularly significant or feature within this book. Sites can be grouped on many different criteria such as by: (1) geographical regions; (2) geological facies in which they are preserved; (3) their age and therefore potential species of track-maker; or (4) by their archaeological or palaeoanthropological significance. While there is a natural tendency to focus on the unusual, biggest, or oldest, in reality footprint sites tend to separate into those which pre-date *Homo sapiens* and those that don't Those that do are limited in number but have the potential to offer information about the evolution of gait between hominin species and as such they accord a level of significance far greater than other footprint sites. Such sites are few in number however and while Holocene sites may not have the glamour of older localities, they have the potential to offer important laboratories in which to explore the interaction of a track-maker's gait with such things as substrate. For ease we have chosen to divide this chapter into those examples that potentially pre-date *Homo sapiens* (Pliocene to Early/Middle Pleistocene) and those that don't (Late Pleistocene to Holocene).

3.1 Pliocene to Early/Middle Pleistocene Tracksites

As we write there are just five human track sites which fall clearly into this time category (Table 3.1; Figs. 3.1 and 3.2), they are:

1. Laetoli (Tanzania) which dates from 3.66 Ma and has been ascribed by most workers to *Australopithecus afarensis*.
2. Ileret (Kenya) which dates from 1.5 Ma and has been tentatively ascribed to *Homo erectus* or potentially to *Paranthropus boisei*.
3. Koobi Fora (Kenya) which dates from just less than 1.4 Ma and has also been tentatively ascribed to *Homo erectus*.
4. Happisburgh (UK) which dates from between 1 and 0.78 Ma forming the oldest hominin track site outside Africa and attributed to *Homo antecessor*.
5. Roccamonfina (Italy) which date from 0.75 Ma and ascribed to *Homo heidelbergensis*.

M.R. Bennett, S.A. Morse, *Human Footprints: Fossilised Locomotion?*,
DOI 10.1007/978-3-319-08572-2_3

Table 3.1 List of documented human track sites

Site #	Site name	Age	Depositional system	Key reference(s)
	Pliocene to early/middle Pleistocene track sites			
1	Laetoli (Tanzania)	3.66 Ma	Volcanic (ash)	Leakey and Hay (1979), Leakey and Harris (1987)
2	Ileret (Kenya)	~1.5 Ma	Fluvial-Lacustrine	Bennett et al. (2009a), Dingwall et al. (2013)
3	Koobi Fora (Kenya)	~1.4 Ma	Fluvial-Lacustrine	Behrensmeyer and Laporte (1981), Bennett et al. (2009a)
4	Happisburgh (UK)	~0.78–1.0 Ma	Fluvial	Ashton et al. (2014)
5	Roccamonfina, Italy	0.385 Ma	Volcanic – Pyroclastic Flow	Mietto et al. (2003)
6	Terra Amata (France)	~0.3–0.4 Ma	Coastal	De Lumley (1966, 1969, 1975), Miskovski (1967)
	Late Pleistocene to Holocene track sites – Europe and Africa			
7	Nahoon (South Africa)	124±4 ka	Coastal Aeolianite	Roberts (2008)
8	Langebaan (South Arica)	~117 ka	Coastal Aeolianite	Roberts (2008)
9	Vartop Cave, Romania	>62 ka	Cave	Onac et al. (2005)
10	Theopetra Cave, Greece	46.327±1.59 ka	Cave	Facorellis et al. (2001)
11	Chauvet Cave, France	~20–30 ka	Cave	Harrington (1999), Garcia (1999)
12	Lascaux, France	Currently uncertain	Cave	Barriere and Sahly (1964), Renfrew and Morley (2009)
13	Niaux, France	~12 ka	Cave	Breuil and Cartailhac (1907), Pales (1976)
14	Grotte Ald`ene, France	8.2±0.013 – 7.79±0.00 ka BP	Cave	Casteret (1948), Ambert et al. (2000)
15	Grotte de Cabrerets (Pech Merle), France	Currently uncertain	Cave	Begouen (1927)
16	Engare Sero (Tanzania)	Currently uncertain	Volcanic Ash	Richmond et al. (2011), Zimmer et al. (2012)
17	Burgos, Spain	~15.6 ka	Cave	Marcos (2001)
18	Tana della Basura, Italy	~12 ka	Cave	Chiapella (1952), Pales (1954, 1960)
19	Tempranas Cave, Spain	Currently uncertain	Cave	Noval Fonseca (2007)
20	Demirköprü, Turkey	~9 ka	Volcanic	Barnaby (1975), Ozansoy (1969), Westaway et al. (2004, 2006)
21	El Azrag, Mauritania	~9 ka	Lacustrine	Mafart (2006)

22	Uskmouth (UK)	6.25 ± 0.08 ka BP	Estuarine	Aldhouse-Green et al. (1992)
23	Kengfig (UK)	3.81 ± 0.04 ka BP – 5.11 ± 0.05 ka BP	Coastal Peat	Bennett et al. (2010)
24	Sefton Coast (UK)	3.649 ± 0.109 ka BP	Coastal	Cowell et al. (1993), Roberts et al. (1996)
25	Walvis Bay (Namibia)	0.5–1.5 ka	Fluvial	Kinahan (1996), Morse et al. (2013)
	Late Pleistocene to Holocene track sites – Americas and Asia			
26	Tibet	~20 ka	Hot sprint, travertine	Zhang and Li (2002)
27	Jeju Island, Korea	Subject to debate	Coastal (Volcanic)	Kim et al. (2010)
28	Willandra, Australia	23–19 ka	Lacustrine	Webb et al. (2006)
29	La Olla, Argentina	~7 ka	Fluvial	Bayón et al.(2012)
30	Monte Hermoso, Argentina	~7 ka	Lagoon	Bayón et al. (2012)
31	Monte Verde II, Chile	~14.6 ka BP	Fluvial	Dillehay (1989, 1997), Meltzer et al. (1997)
32	El Salvador	~0.2–0.8 ka		Haberland and Grebe (1957)
33	Acahualinca	2.1 ka	Volcanic	Williams (1952), Schmincke et al. (2009, 2010)
34	Cuatro Ciénegas (Mexico)	10.55 ± 0.03 ka – 7.24 ± 0.013 ka	Tufa	Felstead et al. (2014)
35	Mud Glyph, Tennessee	<1.5 ka	Cave	Watson (1986)
36	Fisher Ridge Cave, TN	~2.7–3.2 ka	Cave	Watson (1982), Willey et al. (2005)
37	Sequoyah Caverns, Alabama	0.64–0.5 ka	Cave	Sneed (1984)
38	Footprint Cave, Virginia	~0.4 ka	Cave	Crothers (1997), Willey et al. (2005)
39	Lon Odell Cave, Missouri	~600 ya	Cave	Beard (1997)
40	Jaguar Cave, Tennessee	~5.5 ka	Cave	Willey et al. (2005)
41	Oro Grande, California	~5 ka	Fluvial	Rector et al. (1984)
42	Third unnamed, Tennessee	5,210–4,830 cal year bp	Cave	Crothers, et al. (2002)
43	Unknown cave, Kentucky	3.67 ± 0.05 ka BP	Cave	Crothers et al. (2002)
44	Kīlauea, Hawaii	~200 ya	Volcanic	Jaggar (1921, 1934), Moniz Nakamura (2009)

The Site # refers to the location as shown in Figs. 3.1 and 3.2. Only the key references for each site are shown and this is not intended as an exhaustive bibliography. Dates are not shown where there is either no information or they are subject to current debate. While this list is fairly comprehensive it may not be completely exhaustive due to the fact that some sites are only known in historic or grey literature. This list draws heavily on that in Lockley et al. (2008)

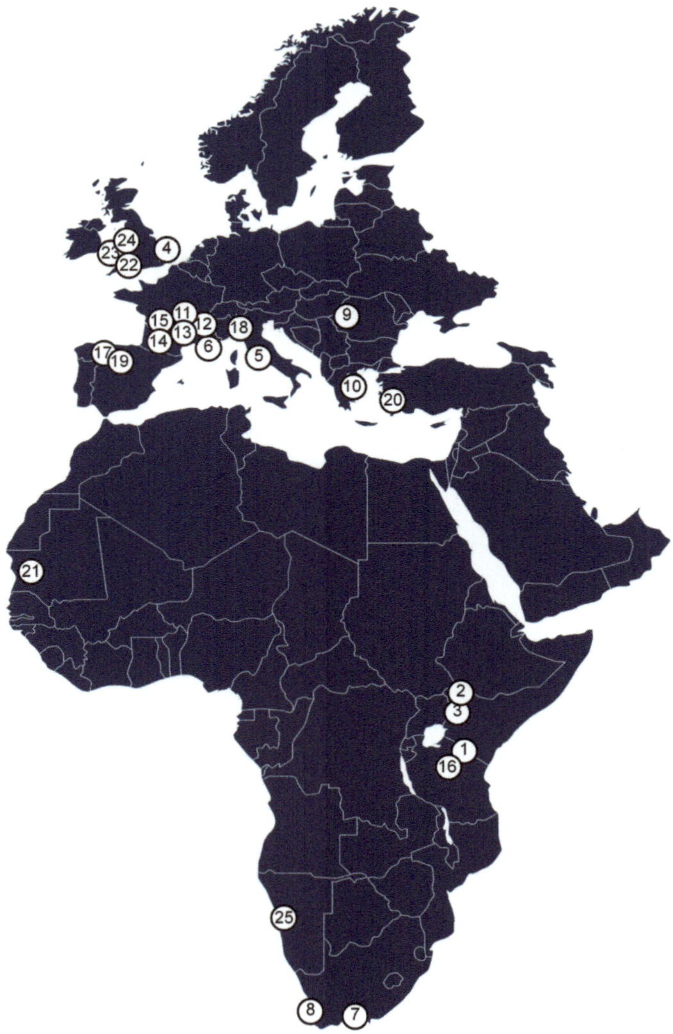

Fig. 3.1 Distribution of human track sites in Africa and Europe. The numbers correspond to those in Table 3.1 where the site details can be found

For completeness it is worth reporting that a single poorly defined track, held as a cast in the Museum of Natural History in Nice and originally taken from the Terra Amata site has also been ascribed to *Homo erectus*, but the contextual information, date and quality of the track limits its value (Fig. 3.3a; De Lumley 1966, 1969; Miskovski 1967; De Lumley et al. 1976). Tracks found during dam construction at Demirköprü in Turkey an example of which is on display at the Museum of Natural History in Stockholm according to Barnaby (1975) and Lockley et al. (2008) were originally assigned to an age of 0.25 Ma by Ozansoy (1969), although are now believed to date from the early Holocene (Westaway et al. 2004, 2006; Lockley et al. 2008).

Fig. 3.2 Distribution of human track sites in the Americas and Asia. The numbers correspond to those in Table 3.1 where the site details can be found

3.1.1 Laetoli Trackways (Tanzania)

The most iconic of all hominin track sites is at Laetoli and was first excavated in the late 1970s and is now dated to 3.66 Ma (Deino 2011) providing one of the earliest direct sources of evidence for hominin bipedalism (Leakey and Hay 1979; Leakey 1981; Leakey and Harris 1987). They have recently been formally defined within systematic ichnotaxonomy by Meldrum et al. (2011; *Praehominipes laetoliensis*) following the lead of Kim et al. (2008).

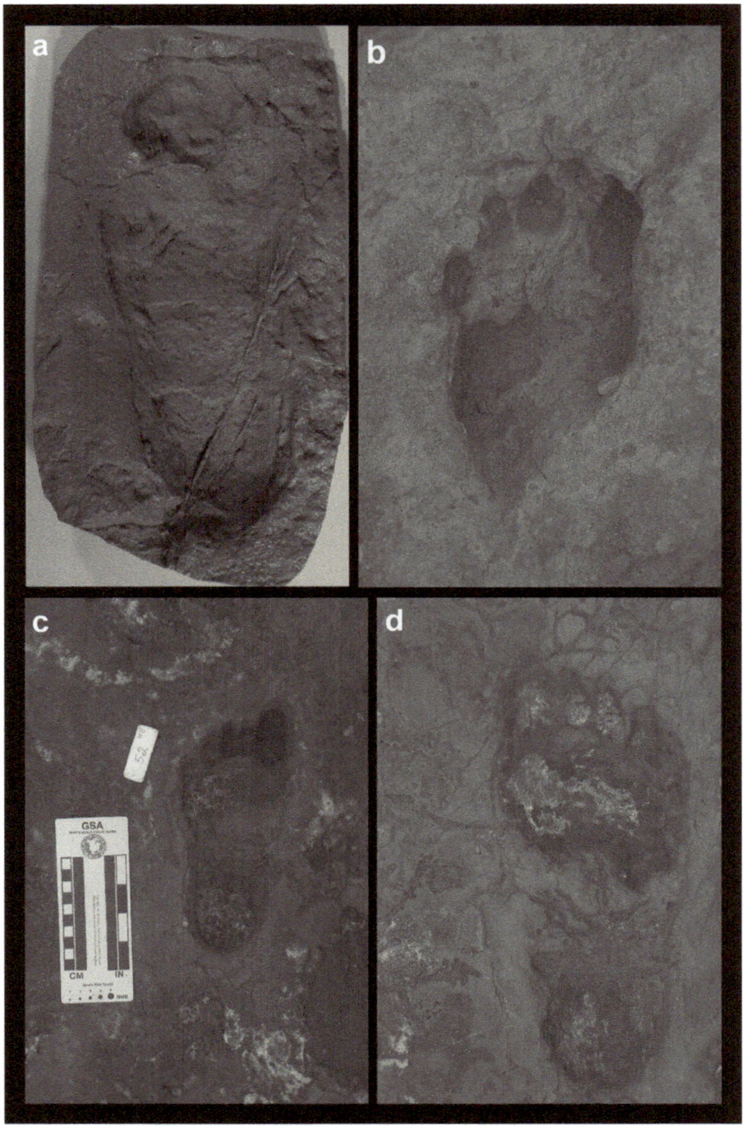

Fig. 3.3 Human tracks. (**a**) Photograph of a cast of the potential *Homo erectus* track at Terra Amata in France. (**b**) Photograph of an original track from Ileret potentially made by *Homo erectus*. (**c**, **d**) Two typical tracks from Walvis Bay (Namibia) showing the quality of preservation

The site lies approximately 36 km south of Olduvai Gorge in northern Tanzania and a total of 18 track sites have been found, of which approximately half have been recorded and only one contains hominin tracks (Musiba et al. 2008). The Laetoli Beds overlie Precambrian basement and can be divided into a lower unit (64 m thick) that consists mainly of air-fall tuffs and water-worked tuffaceous sediments and an upper unit (44–59 m thick) of air-fall tuffs (Drake and Curtis 1987; Hay 1987; Ditchfield and Harrison 2011). The famous Footprint Tuff bearing the hominin tracks (Leakey and Hay 1979; Leakey and Harris 1987) is found in the upper unit. The likely source for the volcanic ash is the extinct Sadiman volcano 20 km to the east. Hay (1987) interpreted the footprint tuffs as having an aeolian origin and suggested that the tephra was deposited over a period of a few weeks at the transition between the dry and wet seasons. The tracks were left almost immediately after rainfall and buried by subsequent ash fall. The distinctive composition of the tephra favoured rapid cementation assisting in the preservation of the tracks (McHenry 2011). According to Lockley et al. (2008) over 9,500 individual animal tracks have been recorded, of which the vast majority are rabbits or other lagomorphs. Other animal tracks include examples of monkeys, antelopes, elephants, rhinos, three-toed horses, cats, hyenas, giraffes, guinea fowl and francolins (Hooijer 1987; Leakey and Harris 1987; Musiba et al. 2008). The main hominin site (Site-G) is approximately 27 m long and consists of three trackways, two of which (G-2 and G-3) are superimposed with a second track-maker (G-3) walking in the footsteps of the first (G-2). The G-1 trail to the west of the double trackway contains 38 prints; there are 31 double prints (i.e. G-2 + G-3). Due to the superimposed nature of the G-2 and G-3 trackways attention has largely focused on the G-1 trackway generating extensive debate and analysis with an ever growing literature (see Meldrum et al. 2011; Fig. 3.4a, b). Figure 3.5 is an optical laser scan of a first generation cast from the National Museums of Kenya showing part of the G-1 and G-2/-3 trackways. The track-maker has been widely attributed to *Australopithecus afarensis* given that small skeletal fragments have been recovered from the Laetoli Beds and it is also the only species of hominin known from the landscape at that time (Suwa 1984; Leakey and Harris 1987; White and Suwa 1987). This view is not shared by all however with some pointing to the possibility of a hitherto un-recorded hominin species as being the potential track maker (Tuttle et al. 1990; Tuttle 2008). White and Suwa (1987) suggest that the track-maker for trail G-1 had a height in the range of 1.1–1.15 m while the G-3 track-maker was slightly taller at 1.32–1.52 m. Tuttle et al. (1990) revised these estimates to 1.22 and 1.44 m respectively based on their modern analogue data. Further discussion about the interpretation of this site can be found in Sects. 6.4 and 6.5.

The sites importance stems from both its antiquity and from the fact that the tracks were made by an extinct species of hominin with a potentially different locomotive style than our own species. In addition the site drove early methodological innovation in the study of vertebrate tracks especially around their recording and interpretation (Leakey and Harris 1987). As is discussed in Sects. 4.3 and 4.4 of the next chapter the conservation of the site has also attracted both research and controversy (Agnew and Demas 1998).

3.1.2 Ileret Footprints (Kenya)

The Koobi Fora Formation, located in the Turkana Basin of northern Kenya, is one of the richest and most famous hominin fossil beds in east Africa (Leakey and Leakey 1978; Harris 1983, 1991). Two sites containing tracks have been excavated in

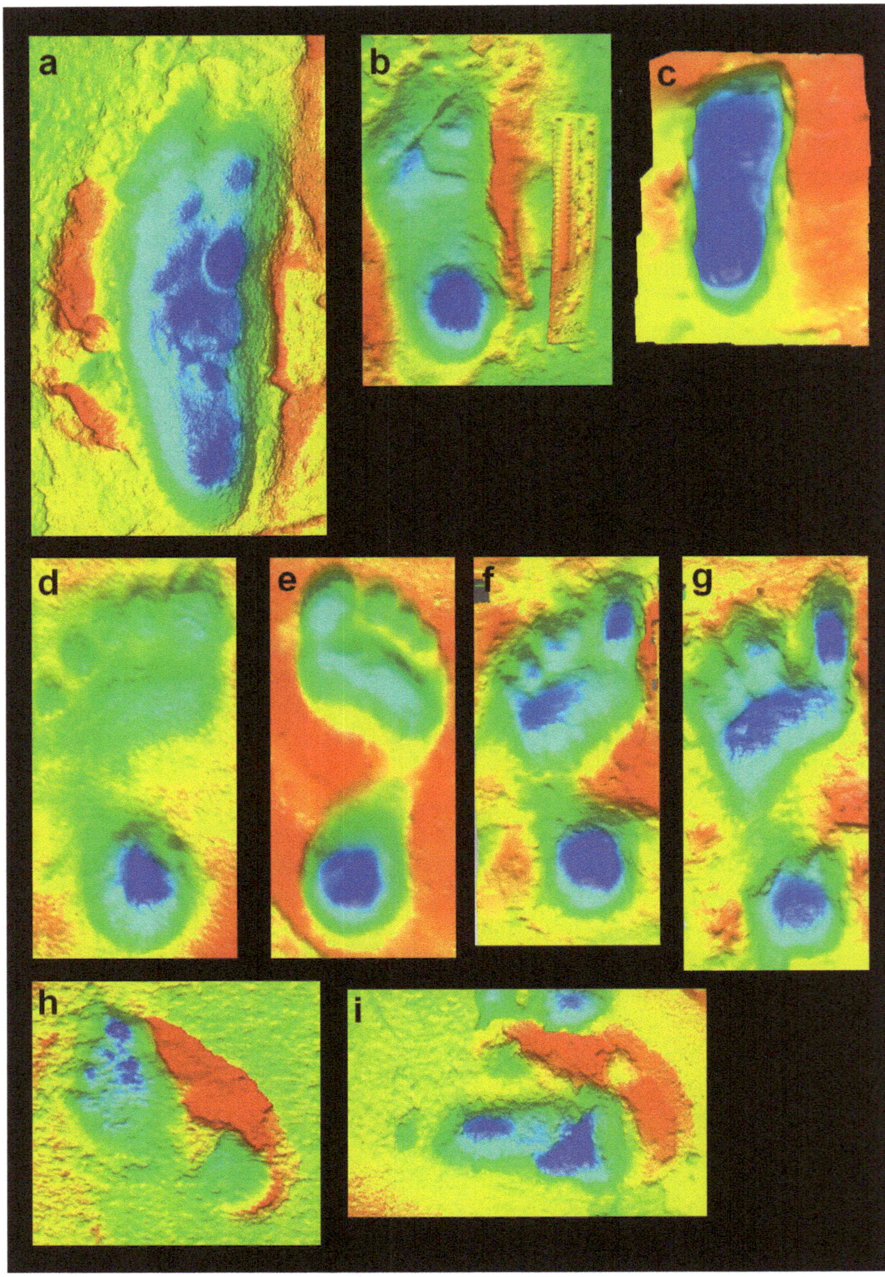

Fig. 3.4 Colour rendered optical laser scans of a range of hominin tracks. (**a**) Track from the G-1 Trail at Laetoli. (**b**) Superimposed tracks from the G-2 and -3 trackways at Laetoli. (**c**) Track from Monte Hermosa in Argentina. (**d, e**) Tracks from Acahualinca in Nicaragua. (**f, g**) Tracks from the Sefton Coast in the UK. (**h**) Track from Nahoon in South Africa. (**i**) Track from Langebaan in South Africa

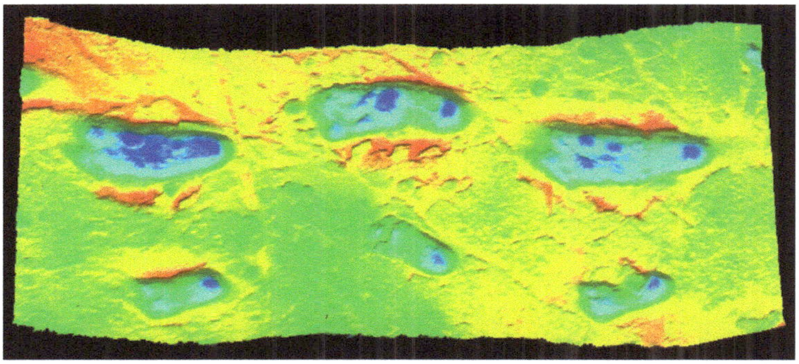

Fig. 3.5 Colour rendered optical laser scan taken from a first generation cast of part of the Laetoli trackways. The scan was made by the authors from material held at the National Museum of Kenya

this basin to date. The older at ~1.52 Ma was reported in 2009 at a site close to the village of Ileret within the Okote Member of the Koobi Fora Formation (FwJj14E; Bennett et al. 2009a). The second site was first described by Behrensmeyer and Laporte (1981) and lies 40 km to the south (see Sect. 3.1.3). FwJj14E consists of an eroding bluff of sediment capped unconformably by Holocene sediments (Galana Boi Formation; Feibel et al. 1989). Excavation at various levels has found multiple track surfaces and is on-going (Figs. 3.3b, 3.6, and 3.7). The surfaces described here are those of 2009 excavations and are located at two stratigraphic levels, with isolated hominin and animal track-bearing strata between. The sedimentary succession consists of over 9 m of fine-grained, normally graded, silt and sand units (0.1–0.5 m) between thicker (0.5–2.0 m) palaeosol units. Isaac and Behrensmeyer (1997) suggest that the sediments around Ileret form part of a low energy fan-delta with numerous seasonally dry distributaries draining into a lake which may have gradually transgressed over at least part of this area. There is no evidence of this transgression at FwJj14E and track-bearing horizons consist of fining-upward waning sheet flood deposits in which coarse sand drapes underlying deposits (and/or the previous flood cycle) fining upwards to fine silts which appear to have been emergent but are not unduly desiccated. They may be representative of either crevasse splays or simply over-bank floods on a low lying flood- or delta-plain. These deposits are inter-bedded with thicker more massive fine sand units which show evidence of palaeosol development. The palaeosols are indicative of wet-dry seasonal conditions as described by Wynn (2004) more generally within the Koobi Fora Formation. Within this succession there are three re-worked volcanic ashes; the upper ash (Northern Ileret Tuff) forms a prominent landscape benchmark that correlates with nearby sites containing traces of hominin activity including cut bones (Pobiner et al. 2008). The ash layers have been correlated geochemically to dated tuffs within the Turkana Basin thereby providing an age of 1.51–1.52 Ma for the upper tuff and 1.53 Ma for the lower tuff (Bennett et al. 2009a).

Dingwall et al. (2013) provides a summary of hominin tracks found as of 2011 (Fig. 3.6). There are two main track bearing horizons; a lower surface consisting

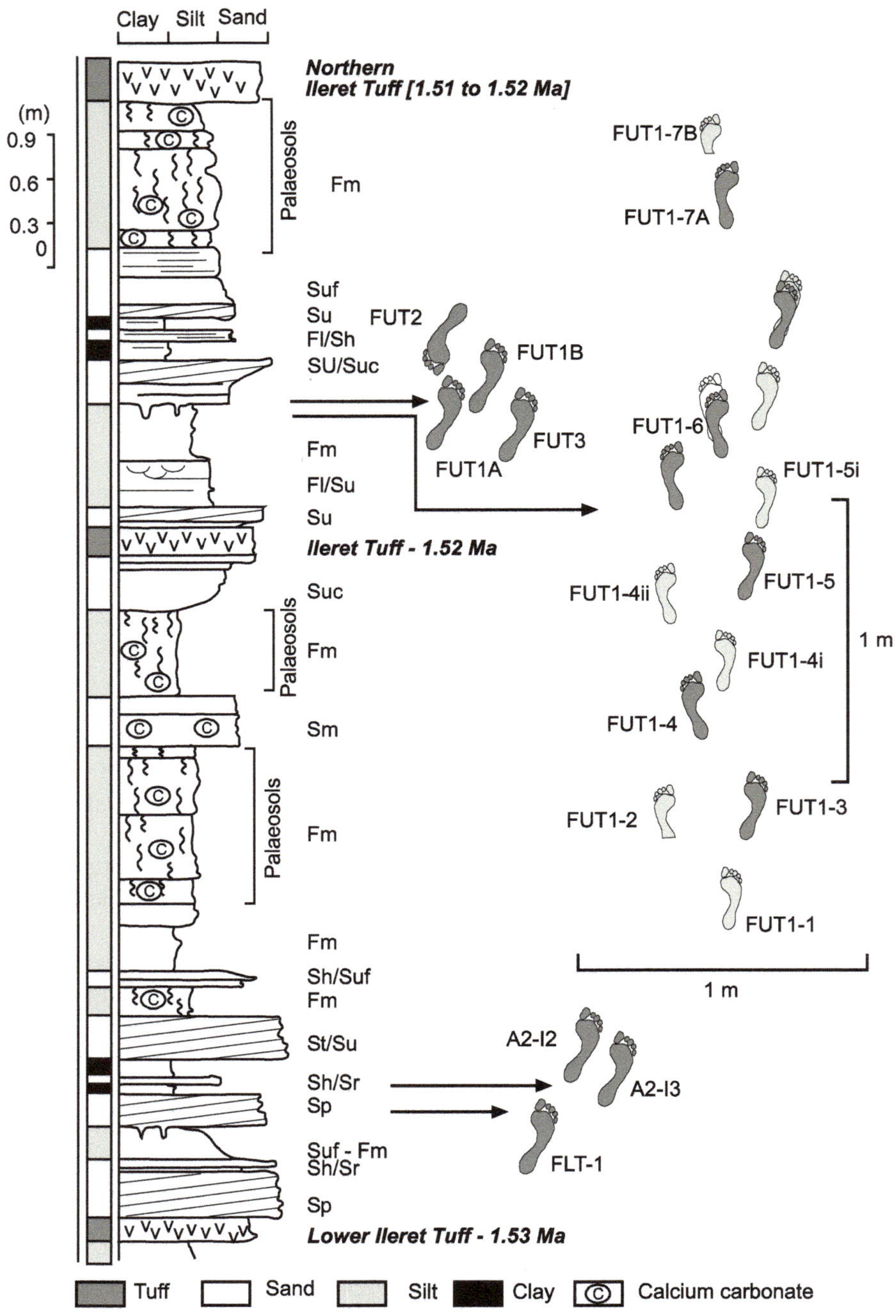

Fig. 3.6 Stratigraphy of the Ileret hominin track site following Bennett et al. (2009a) and the interpretation of the track sequence following Dingwall et al. (2013)

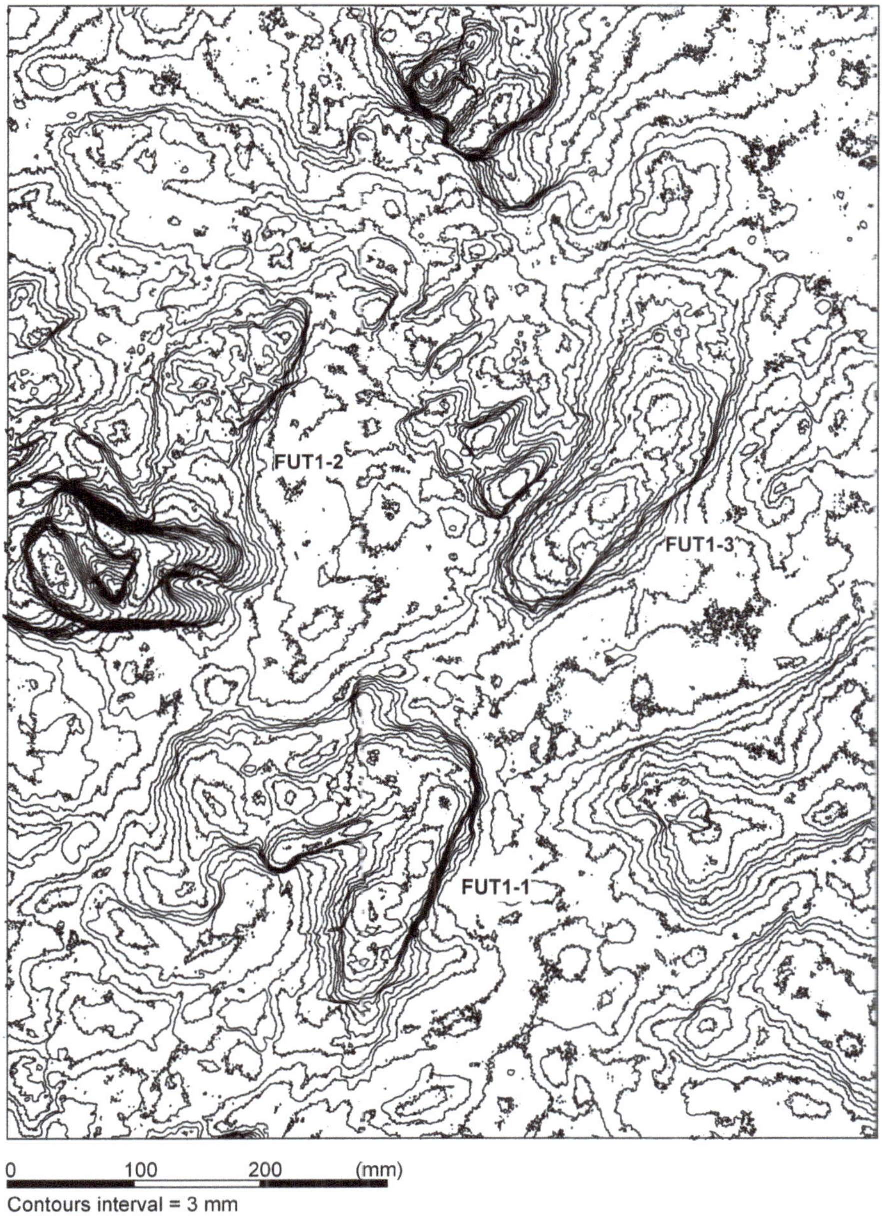

Fig. 3.7 Contour map of part of the longest trackway of tracks on the Upper Footprint surface at Ileret

of up to three tracks and immediately above are two additional prints in thin and heavily trampled areas with tracks penetrating one or more silt horizons. The upper surface consists of a number of isolated prints and one short trail of nine prints which is interpreted by Dingwall et al. (2013) as being made by two individuals travelling in a similar direction (Fig. 3.6). All the human tracks occur in association with a rich record of quadrupedal mammals and birds. Dingwall et al. (2013) estimate walking speeds of between 0.45 and 2.2 ms^{-1} made by heavy (41.5–60.3 kg), tall individuals (1.526–1.858 m; see Sects. 6.2 and 6.4). The prints were tentatively attributed by Bennett et al. (2009a) to *Homo erectus* although Dingwall et al. (2013) has suggested that they could have been made by a male *Paranthropus boisei*.

The sites importance again stems from both its antiquity and the fact that the tracks were made by an extinct species of hominin. Comparison of track morphology and inferred locomotive styles between this site and that at Laetoli has the potential to provide information about the way in which hominin locomotion may have changed (or has not changed) across the major evolutionary transition from the habitual bipeds of the *Australopithecus* genus to the endurance walkers and runners which characterise more modern species such as *Homo erectus* and ourselves. This is a subject that is picked up again in Sects. 6.4 and 6.5.

3.1.3 Koobi Fora Footprints (Kenya)

Approximately 45 km to the south of Ileret there is a second footprint site (GaJi10) first reported by Behrensmeyer and Laporte in 1981 consisting of a single trackway of poorly defined prints which in contrast to those at Ileret, and for that matter most documented hominin track sites, are believed to have been imprinted subaqueously (Laporte and Behrensmeyer 1980). The footprint surface occurs below a prominent tuff, sampled and correlated by Bennett et al. (2009a) to the Akait Tuff dated to 1.435 Ma (Brown et al. 2006). Re-excavation of these prints by Bennett et al. (2009a) uncovered four of the original seven prints. The lithofacies around GaJi10 is consistent with a low energy fluvial-lacustrine system subject to both short-term seasonal and millennial-scale water variations (Behrensmeyer 1975; Lepre et al. 2007). This landscape was rich in a diverse range of vertebrate and semi-aquatic fauna and has yielded a plethora of vertebrate remains with aquatic and semi-aquatic fauna being more common around GaJi10 (Behrensmeyer 1975; Bennett et al. 2014a).

The original surface excavated by Behrensmeyer and Laporte (1981) contains over 89 distinct impressions (c. 12 m^2) identified as the tracks of large vertebrates (hippopotami) in addition to a short hominin trackway. According to Behrensmeyer and Laporte (1981) the site was covered by shallow water and interpretation based in part on the presence of a wading bird track, although it is possible that the hominin trackway was made at a subsequent lake low-stand. Behrensmeyer and Laporte (1981) attributed the tracks to *Homo erectus*, an interpretation supported by Bennett et al. (2009a) upon re-excavation. Track anatomy is poor compared to the prints at

Ileret perhaps reflecting the sub-aqueous conditions. A second excavation on the same palaeo-surface was conducted by the authors 80 m to the south. This much larger excavation (13 m along strike and 3 m wide) does not contain any visible hominin tracks, but does contain over 240 individual tracks interpreted by as being formed by swimming hippopotami 'punting' or bottom-walking along the bed of a shallow water body (Bennett et al. 2014a). The depth of this water body is estimated at between 0.5 and 1.5 m and is a deeper water equivalent to that found in the excavation further north in which the hippopotami tracks were formed by normal walking as the water body is too shallow to allow swimming. This animal assemblage provides a sharp contrast to that described at Ileret which is subaerial and dominated by bovid, suid and equid tracks.

The sites importance again stems from both its antiquity and the fact that the tracks were made by an extinct species of hominin. Track topology is anatomically poor compared to that at Ileret and the site does not have much therefore to contribute to the discussion on the evolution of locomotive styles. It does however provide evidence of hominin exploitation of a different habitat to that found at Ileret and is one of a very few, potentially only, recorded instances of track preservation in a subaqueous environment. The association of the hominin tracks with aquatic animal tracks is also unique and important to the study of swim tracks (Bennett et al. 2014a). Finally the site illustrates the potential of the Koobi Fora Formation and similar fluvial-lacustrine systems to preserve human and animal tracks the implication being that there are probably lots more tracks to be found in this region in the future.

3.1.4 Happisburgh (United Kingdom)

Ashton et al. (2014) have recently reported hominin tracks exposed on the Norfolk coast which may provide the oldest known hominin track site outside of Africa a title held previously by those of Roccamonfina. At Happisburgh the Cromer Forest-bed Formation dates from 2.0 to 0.5 Ma and is characterised by a series of estuarine, fluvial and alluvial sediments that interdigitate with near-shore marine sediments. This formation is associated with lower Palaeolithic archaeology at Pakefield dating to 0.7 Ma (Parfitt et al. 2005) and at Happisburgh to between 0.85 and 0.9 Ma (Parfitt et al. 2010). The formation is exposed by coastal cliff erosion at Happisburgh where a palaeo-surface with potential hominin tracks has been found, although subsequently the tracks have been lost to erosion (Ashton et al. 2014). A total of 152 potential tracks were recorded with lengths between 160 and 172 mm and widths of 80–27 mm. Ashton et al. (2014) argued for a hominin track-maker on the basis of track size and elongation, as well as several poorly defined trackways. It has to be acknowledged however that there is little anatomical detail to verify this and definitive trackways are poorly developed. Ashton et al. (2014) estimate stature to be between 0.93 and 1.73 m and accordingly attribute the track-maker to *Homo antecessor*.

This site is important due to its antiquity and location outside Africa and it provides a tantalising glimpse of what may be revealed by future coastal erosion. The current absence of any good digital elevation models for the tracks coupled with their poor preservation of anatomical detail precludes much more than the most basic analysis (Ashton et al. 2014), but hopefully if more tracks are uncovered in the future by continued coastal erosion of the beds and captured digitally the site will provide material that will allow comparison with other tracks made by extinct hominins.

3.1.5 Roccamonfina (Italy)

The Roccamonfina site is located in the municipalities of Tora and Piccilli in southern Italy. Three human trackways descend a steep (30–80°) primary slope formed by the surface of a pyroclastic flow (Mietto et al. 2003; Avanzini et al. 2008). The site has been dated by Ar/Ar to 345±6 ka BP (Scaillet et al. 2008) and the track-maker has been tentatively ascribed to *Homo heidelbergensis*. Oblique step lengths of 600 mm, stride lengths of 1.2 m and individual track lengths of 240 mm are typical of this unusual site. Trackway-A is 13.4 m long and consists of 27 left-right patterned prints descending a vertical distance down the flow surface of 4.26 m. The trackway has a Z-shaped path with two sharp turns indicative of an individual making a tentative and careful decent of a steep, partially unstable slope (Fig. 3.8). Trackway-B is 8.6 m long and consists of 19 tracks descending 2.91 m in a single curved decent line with frequent slips and irregularities suggesting a less careful descent. The third trackway (Trackway-C) is less well developed and follows a straight-line of 9.98 m, with ten tracks descending 2.56 m in elevation. Anatomical preservation in all cases is poor due to: (1) the unstable substrate; (2) the steep surface slope down which the track-maker was descending; and (3) the unusual pattern of gait associated with both of the above. Avanzini et al. (2008) estimate that the height of the track-maker was approximately 1.56 m and perhaps travelling at a speed of 1.09 ms^{-1}.

The primary importance of this site was, until recently, that it was Europe's oldest footprint site. However its true claim to fame lies not in its antiquity, but in the unique nature of the preservation and the tracks themselves. Most track sites consist of a horizontal surface, this site contains tracks that descend a steep slope and are particularly emotive since they capture a sense of movement and haste of the ancient track-maker even if they yield little about the anatomy of the track-makers foot or of their normal pattern of gait. The site is a source, quite rightly, of immense local civic pride and has a unique and special place in the record of human track sites.

3.2 Late Pleistocene to Holocene Tracksites

Within the published record as of April 2014 there are well over 30 Late Pleistocene or Holocene human track sites (Table 3.1; Lockley et al. 2008). We have selected just a few from this list to represent a range of different depositional settings.

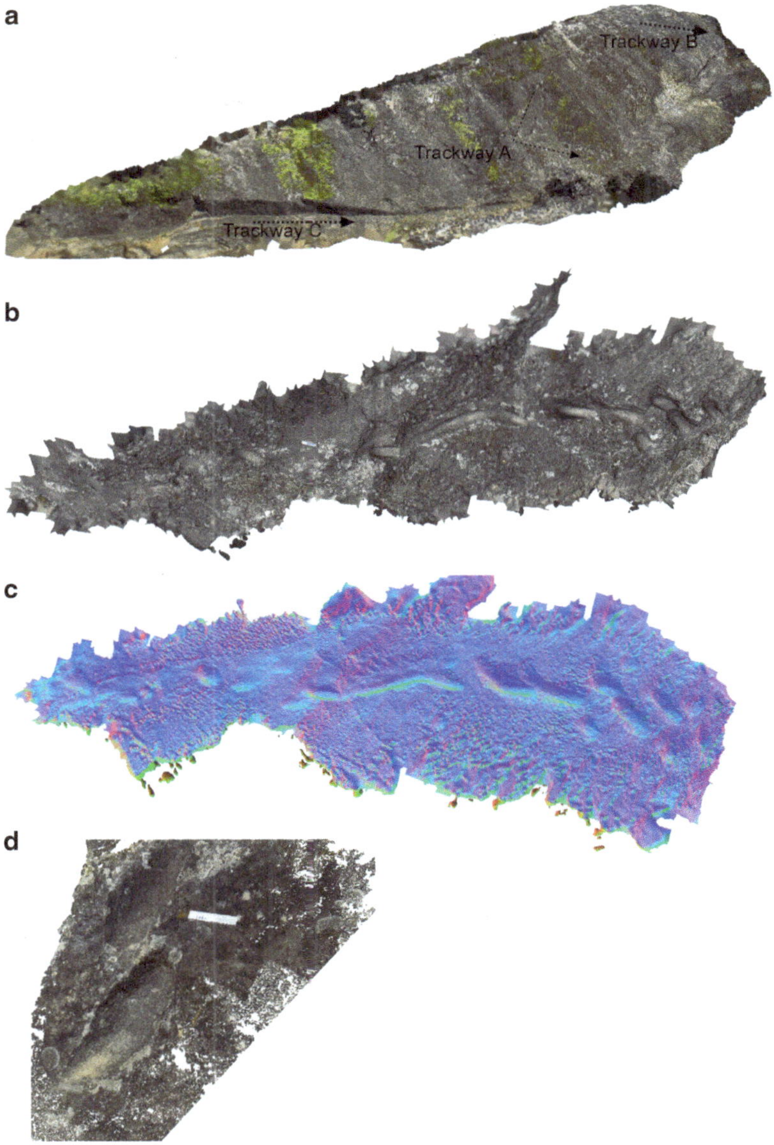

Fig. 3.8 The Devil's Tracks at Roccomonfina in Italy. Unpublished three dimensional models made via photogrammetry by Peter Falkingham. (**a**) Over view of the site showing the palaeo-slope. (**b, c**) Trackway B showing the slippage which occurred as the track-maker negotiated the steep slope. (**d**) Close-up of two tracks in Trackway A

3.2.1 *Acahualinca (Nicaragua)*

This is one of the most spectacular human track sites found to date with tracks preserved in fine volcanic ash dating to around 2,120 ± 120 BP (Brinton 1887; Brown 1947; Bryan 1973; Lockley et al. 2007, 2009; Schmincke et al. 2009, 2010; Fig. 3.4d, e). The footprint horizon is composed of a lower 5–15 cm thick coarse-grained vesicle tuff capped by a medium to fine-grained tuff up to 3 cm thick derived from a hydroclastic eruption of the nearby Masaya Volcano 20 km to the south (Schmincke et al. 2009). A series of sub-parallel trackways were made by a group of 15–16 individuals who according to Schmincke et al. (2010) were trying to escape the eruptions which would have been particularly violent characterised by lighting, clouds of steam and damp ash falls. Using track length and stride lengths Schmincke et al. (2010) suggested that the group was mixed with males walking on the flanks of a central group of women and children. The quality of track preservation is exceptional and many of the tracks show well-developed rim structures, with superimposed tracks (overprinting) common. Due to the exceptional preservation and the fact that the site is preserved in a purpose built museum the tracks where used by Kim et al. (2008) to define the ichnotaxa for modern human tracks. This track site is the only site that the authors are aware of that is currently conserved within a purpose built shelter and as such is of particular note (see Sect. 4.4), it is also afforded a particular status due to it being the reference site for the tracks made by modern human ichnotax. The paper by Schmincke et al. (2010) illustrates some of the challenges and pitfalls in the interpretation of human track sites and the temptation to over interpret a series of tracks in light of other information. Despite this the paper is an excellent example of its type and illustrates the challenges associated with deducing the number of track-makers from a series of closely superimposed trackways. It is one of two potential sites that record a human exodus in the face of a volcanic eruption, the other site being that of Avellino in Italy which is discussed later in Sect. 3.2.7.

3.2.2 *Sefton Coast (United Kingdom)*

Of the track sites exposed around the coast of the UK (Uskmouth, Severn Estuary, Aldhouse-Green et al. 1992; Kenfig, South Wales Bennett et al. 2010) perhaps the best known are those exposed on the Sefton Coast near the town of Formby just outside Liverpool (Roberts et al. 1996; Huddart et al. 1999a, b). In excess of 145 human trackways dating to between 3,230 and 3,649 BP have been recovered on the foreshore as autumnal storms draw down the beach (Roberts et al. 1996; Fig. 3.4f, g). A series of potentially older tracks have also been recorded although the age remains uncertain, but potentially between 8.8 and 6 ka according Gonzalez and Huddart (2002). A combination of stratigraphic and palaeoenvironmental analysis has provided a complex picture of coastal evolution in this area dominated by a sustained transgression of sea level between 7 and 5.6 ka in which a coastal barrier moved onshore creating a laterally variable pattern of lagoons, salt marsh and mudflats in association with dune and near-shore sediments. Human and animal tracks have been found in varying combination along the Formby sea front between Dale Slack Gutter in the north and Lifeboat Road in the south as these sediments are revealed in the current foreshore by coastal erosion. Animal tracks include those of deer, crane, unshod horses, domesticated ox and auroch. Human tracks range in size between 1.45 and 1.66 m the former being ascribed to females. Roberts et al. (1996) suggests that the larger tracks (males) are associated with above-average speeds and occur commonly with deer tracks, while the smaller (females and children) are associated with lower speeds and therefore a different type of activity. The methods by which the speed estimates were made, and the sex of the track-maker determined, are not clear in the published papers and while the conclusions may be correct it does not provide an ideal role model for track studies. There are also assertions made in the literature (e.g., Roberts et al. 1996; Gonzalez and Huddart 2002) about the presence of pregnant track-makers inferred from the tracks but on what basis is not clear, assertions around the presence of deformities such as fused or missing toes, congenital bunions and other similar pathologies are confirmed however by the authors own observations at the site. A full review of the site and the palaeoenvironmental context can be obtained from Huddart et al. (1999a, b) and Gonzalez and Huddart (2002). Despite reservations about the sophistication with which the tracks have been investigated in the past, what is clear is that they are numerous and the site still provides a potentially large track resource. Despite the best efforts of the amateur enthusiast Gordon Roberts who has worked tirelessly to promote the site to the scientific community it is a site whose full potential has yet to be seen and is one worthy of continued investigation.

3.2.3 *Walvis Bay (Namibia)*

South of Walvis Bay in the northern margins of the Namib Sand Sea the ephemeral Kuiseb River periodically drains into the Atlantic during flood events. Large dunes moving over surfaces of silt and sand deposited during flood drainage form an area referred to as the Kuiseb Delta, which has a rich Pre-Colonial archaeological record associated with a complex transhumant land-use system that combined seasonal inland pastures with exploitation of shellfish and other coastal resources (Kinahan 1996, 2001). Animal and human tracks are present at several locations preserved on the inter-dune mudflats (Kinahan et al. 1991; Kinahan 1996; Kinahan 2013; Morse et al. 2013; Figs. 3.3c, d, 3.9). Recent dating of these tracksites has shown the inter-dune mudflats are diachronous surfaces and various track sites have yielded dates of between 500 and 1500 years BP (Fig. 2.8). Morse et al. (2013) described an extensive area of human and animal tracks at one site using the variation in track morphology along single trackways to examine the influence of substrate. The value of such sites is not in their archaeological or anthropological significance, although this is of local importance, but that the sheer number of tracks provides a natural laboratory in which to explore the formation and preservation of tracks and the morphological variation associated with a range of track making variables.

A second site close to that documented by Morse et al. (2013) contains a spectacular series of tracks made by children (Fig. 3.9; Bennett et al. 2014b). The site is heavily desiccated and contains a series of short trackways indicating the penecontemporaneous movement of a small group of individuals from the south-east to the north-west across the site. The human tracks cross-cut and post-date a series of animal tracks made by small bovids probably sheep/goat (ovicaprids) which appear to have been moving as a loose flock in front of the human track-makers. While the overall direction of movement of the human group traversing the site is apparent, individual prints are difficult to attribute to single trails, which reflects: (1) a lack of linearity to some of the trails, with abrupt changes of direction and in some case short discontinuous print sequences oriented against the general direction of movement trend, which might suggest a group of individuals moving in a common direction but in an irregular almost 'playful' fashion; (2) that prints are frequently missing suggesting a lack of continuous preservation primarily due to the shallow nature of many of the prints; and (3) deep desiccation cracks, which occur ubiquitously across the site. Taking the 31 most complete, (as opposed to partial) prints it is possible to examine the distribution of sizes and to infer a group of between six to nine individuals. Average track lengths (Pternion-second toe) vary from as little as 114–206 mm with a mean of just 152 mm. This site is discussed further in Sect. 6.2 in connection with the inference of age from track lengths, but the implication from these track lengths is that some of them may have been made by children as young as five years old if not younger.

3.2.4 Monte Hermoso (Argentina)

Along the south eastern coast of the Buenos Aires province deposits of Late Pleistocene and Holocene age outcrop within the current littoral zone dating from between 16 and 4.8 ka BP (Aramayo et al. 2005; Bayón et al. 2011). These deposits are around 4 m above current sea level close to Pehuen-Có decreasing in elevation toward the east and Monte Hermoso. These deposits represent a transition from continental to marine environments with wide floodplains and interconnected ponds between 16 and 12 ka BP, being replaced by dunes and interconnected shallow pools between 8.8 and 7.1 ka BP, after 6.9 ka BP the marine influence increased and the area was finally transgressed by the sea between 5.3 and 4.8 ka BP. These deposits contain a rich and important archaeological and palaeontological record which includes animal and human footprints (Aramayo 2009; Aramayo and Manera de Bianco 2009). This whole area forms part of the Pehuen Có Palaeoichnological Preserve.

The site at Monte Hermoso contains numerous bird and mammal footprints including a significant number of human tracks (Bayon and Politis 1996; Aramayo 2009). For example, within one area of 438 m^2 472 human tracks, 35 bird and two even-toed ungulated mammal prints have been recorded. On the basis of foot lengths

Fig. 3.9 Human tracks from a site close to Walvis Bay, Namibia. On the basis of size these tracks are believed to have been made by very young children, potentially as young as 5 years or less (Bennett et al. 2014b)

Fig. 3.10 Tracks at Homonso (Argentina) exposed just below the current beach, but formed originally in fluvial and lagoon sediments

most tracks appear to have been made by women and children rather than by males who appear to have not frequented the area (Aramayo 2009). The tracks formed around both fresh water pools and estuarine lagoons and the deposits have been dated to 7,125 BP (Aramayo 2009). In conjunction with other archaeological evidence a picture emerges of sites revisited on multiple occasions by hunter-gatherers of all ages during a period lasting at least hundreds of years (Bayón et al. 2012; Blasi et al. 2013). A series of tracks was excavated by one of the authors at Monte Hermoso in December 2012 are illustrated in Fig. 3.10. Isolated human tracks have also been reported at Pehuen-Co and may date from as early as 12,000

BP providing potentially some of the earliest evidence for human colonisation of South America (Aramayo and Menara de Bianco 2009; Bayón et al. 2011). These sites are of immense archaeological importance within South America, but also illustrate the complex, laterally variable and diachronous nature of many tracked surfaces. The palaeoenvironments recorded in these deposits changes laterally rapidly as does their age, posing considerable challenge in obtained reliable dates. Moreover the occurrence of these immensely important deposits, in some cases, within a few centimetres of the contemporary beach surface poses a considerable conservation challenge.

3.2.5 *Cuatro Ciénegas (Mexico)*

The Cuatro Ciénegas Basin in northern Mexico provides a very distinctive geoarchaeological environment characterised by more than 200 carbonate-rich pools (pozas) fed by groundwater springs around the margins of which there extensive deposits of tufa and travertine (Gonzalez et al. 2007). There is a rich archaeological record of hunting, gathering and fishing associated with the occupation of caves around the basin margins (Felstead et al. 2014). Two prints from this site were recovered in the early 1960s and are on display at the Museo del Desierto in Saltillo (Coahuila), but without any locality information (Gonzalez et al. 2007). In 2008 the original site of these tracks was apparently re-discovered and documented (Gonzalez et al. 2009; Bennett et al. 2009b). Figure 3.11 shows a single trail of 5 complete tracks; other isolated tracks were recorded at the site as well (Gonzalez et al. 2009). A new U-series date for tracks in the museum of 10.55 ± 0.03 ka was reported by Felstead et al. (2014), which is older than the in situ tracks dated to 7.24 ± 0.13 ka. While the assumption is that the museum specimens come from the same site this may not actually be the case given the dating and it illustrates the challenges of conservation by block removal, something which is explored further in Sect. 4.3. The museum specimens are the oldest known human tracks in Mexico. The other point of note about this site is that the tracks are found in tufa which is a relatively unusual preservation scenario within the literature despite the high number of cave sites, with only one other example from Tibet known to the authors (Zhang and Li 2002).

3.2.6 *Cave Sites and the Jaguar Caves (Tennessee, USA)*

Within the published record there are a large number of cave sites containing preserved human tracks especially within Europe (Table 3.1). The oldest currently known cave site with human tracks in Europe is Vârtop Cave in Romania, which has been dated to approximately 62 ka (Onac et al. 2005). Those described by Facorellis et al. (2001) from the Theopetra Cave in Greece may also pre-date the last glacial maximum with an age of around 43 ka BP. There are numerous examples in France which occur with cave art and extensive evidence of cave occupation. Tracks are

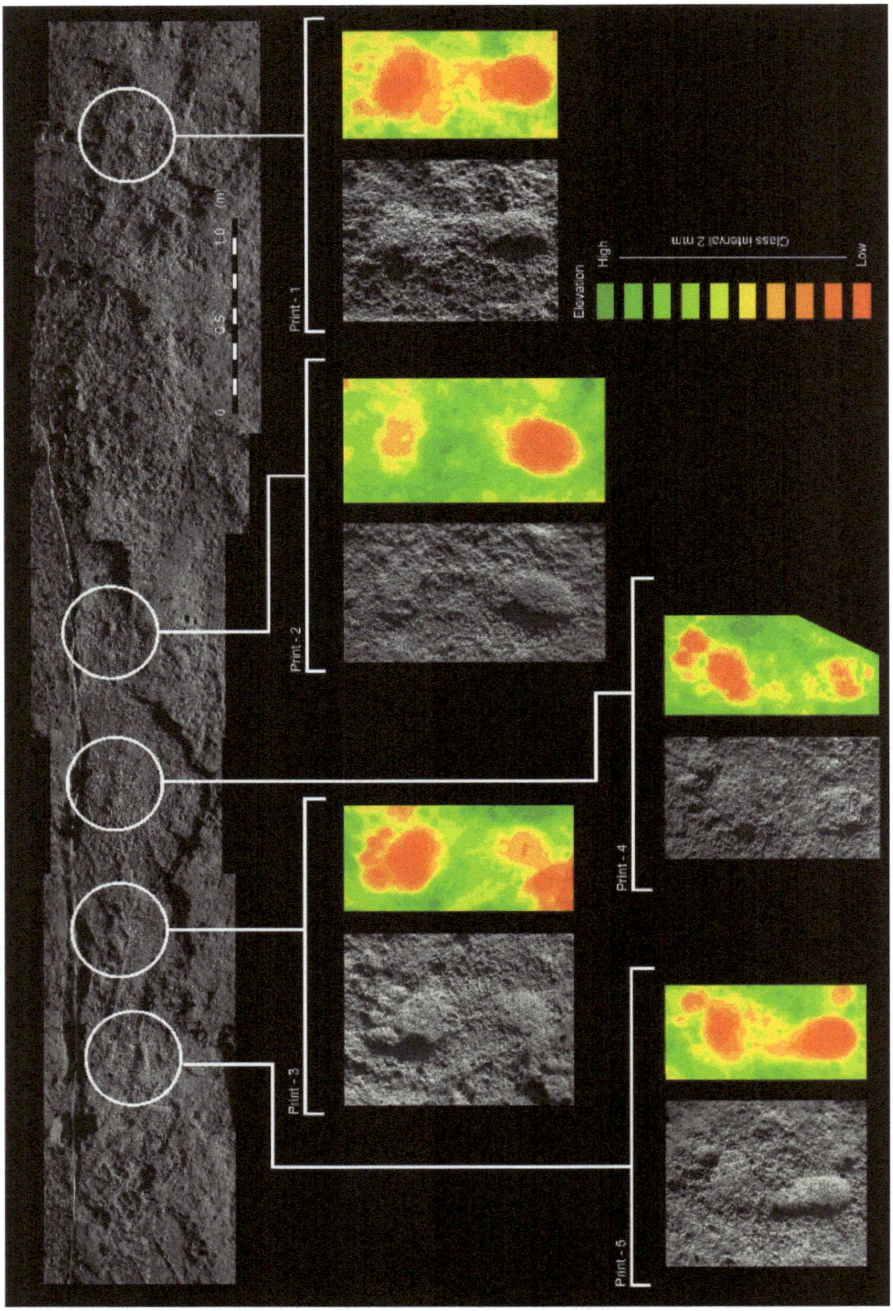

Fig. 3.11 Tracks at Cuatro Ciénega in northern Mexico. These tracks are unusual in that they are preserved in tufa. See Bennett et al. (2009b) and Felstead et al. (2014) for details

known from the caves of Lascaux (Barriere and Sahly 1964), Niaux (Pales 1976), Grotte Aldène (Casteret 1948; Ambert et al. 2000), Chauvet (Harrington 1999; Garcia 1999, 2001; Valladas et al. 2005) and Grotto de Cabrerets (Pech Merle; Begouen 1927; Vallois 1927, 1931). Most of these tracks have been assigned to *Homo sapiens* and date to the late Pleistocene or early Holocene. Those at Tana della Basura in northern Italy have been attributed to *Homo neanderthalensis* (Chiapella 1952; Pales 1954, 1960, although in practice may be too young given new dating (Molleson et al. 1972; Onac et al. 2005). There are numerous other European cave examples with new one emerging from time to time, such as Ciur-Izbuc Cave in Romania.

In illustration of this type of environment however we draw on an example from the Americas and the tracks found in the Jaguar Caves of Tennessee (Watson et al. 2005; Willey et al. 2005, 2009; Lockley et al. 2008). The tracks were discovered by modern cavers in 1976 who were exploring new passages of the cave system in Mississippian and Pennsylvania Limestones of the Cumberland Plateau which extends for over 13 km. The cave gets its name from jaguar (*Panthera onca*) tracks which are found in several mud-floored passages (Watson et al. 2005). The animals became trapped in the cave system on at least two separate occasions between 10 and 35 k BP. These tracks are temporally and spatially separated from the later human tracks. Fossil remains of a wide variety of other species have also been recovered from the caves (Watson et al. 2005). These include the following extinct taxa: passenger pigeon (*Ecotopistes migratorius*), mastodon (*Mammut americanum*), long nosed peccary (*Mylohyus nasutus*), dire wolf (*Canis dirus*), horse (*Equus*), tapir (*Tapirus*), and camel (*Camelops*). The human tracks have been dated by the association of pieces of burnt charcoal – assumed to be from torches carried by the pre-historic cavers and give an approximate age of 4.5–5 k BP. For example, charcoal collected from a dry passage between the Only Crawl and Tremendous Trunk yielded calibrated dates of 5465–4870 years BP and 5,600–5,090 years BP (Robbins et al. 1981). A third charcoal sample was collected from Aborigine Avenue and dated 5,575–4,990 BP. While these dates provide evidence of the presence of humans carrying torches they do not in themselves provide a date for the tracks, although in practice it is hard to see how the track-maker could have walked safely in the dark! Nearly 300 tracks have been found in soft, moist mud in a side passage called Aborigine Avenue. Tracks are absent on associated flow stones and tufas, although water dripping from the ceiling has pocketed in a few of the prints. In general the tracks are largely untouched due to the remoteness of the passages and the moisture and relative humidity has apparently remained largely unchanged since the formation of the tracks. According to Robbins et al. (1981) at least some of the track-makers where shod with thin flexible footwear, perhaps woven from plant fibres like the shoes lost or discarded by aboriginal cavers in the Mammoth Cave System (Watson 1969, 1974). Since the prints' discovery, their destruction has accelerated due to the passage of modern cavers and archaeologists. The very process of observation is in itself destructive, for example on the discovery trip modern cavers walked over some of the prehistoric tracks before recognizing them. The site illustrates the challenge posed by preserving tracks in caves such as this, but also the potential of such environments to record pristine tracks. From a forensic perspective it is worth reflecting on the fact that the process of observation may in fact alter the nature of the record.

3.2.7 Other Notable Sites

There are a number of other tracksites which are worth drawing the reader's attention to, at least briefly. The human tracks on Jeju Island in South Korea (Kim et al. 2009), 90 km south of the Korean Peninsula, have attracted some controversy with respect to their age which has variously been placed at late Pleistocene (c. 19–25 ka BP) or late Holocene (c. 3.7 ka BP) and is a subject of on-going debate. The volcanic complex which forms the island has been emplaced over the last million years and the tracks are found in re-worked volcaniclastic sediments, the lithostratigraphic correlation of these sediments, and therefore their age is a source of differing opinion with workers suggesting that they pre-date (Kim et al. 2010) or post-date the Songaksan Tuff which has been dated on the basis of OSL to around 7.0 ± 0.3 ka (Cheong et al. 2007). This debate is one of local geo-politics, but is also spurred on by the relative age of these tracks with respect to the only other Asian tracksite which is in Tibet. In Tibet hand and footprints were found in travertine and have been dated by Zhang and Li (2002) to around 22 ka BP. Irrespective of the age of the Jeju tracks they provide an impressive series of human tracks numbering over 500, typically 120–260 mm long and 60–120 mm wide and being recorded in at least nine trackways, along with tracks of artiodactyla, proboscidea and a diverse range of birds (Kim et al. 2009, 2010).

Other impressive tracksites in volcanic ash include those described by Mastrolorenzo et al. (2006) associated with the 3,780 ka BP Plinian eruption of Vesuvius where an early violent pumice rich ash fall was followed by later pyroclastic surges covering an area over 25 km from the volcano. Footprints reveal a sudden en mass evacuation of thousands of people at the start of the eruption, most of who appear to have survived unlike the later eruptions at Pompeii. Di Vito et al. (2009) described the destruction and abandonment of the Bronze Age village of Afragola and in particular documents the tracks associated with each layer and event of the eruption. The successful escape of the entire population is apparent from the lack of human remains and from the literally thousands of human tracks on the surface of the deposits which record the early phases of the eruption. Rapid cooling of these deposits must have occurred to allow people to move across them barefoot, probably due to their thinness and presence of abundant water vapour. The tracks vary from single trackways made by solitary individuals, to confused 'pathways' caused by multiple individuals moving at once along a similar path. The presence of longitudinal grooves in association with some of these trackways may be indicative of individuals dragging possessions and heavy objects with them. Tracks occur on multiple horizons indicating significant breaks between the emplacement of the different pyroclastic flows and the on-going evacuation of individuals during the eruption. The sheer numbers of tracks available at these sites, archaeological excavations which have now been back filled, provide a tantalising glimpse at a resource which could be used to reconstruct the biometrics of an entire village population. Such a study would be a very valuable addition to the study of human tracks.

In a succession of papers Meldrum (2000, 2002, 2004a, b, 2007; Meldrum et al. 2011) has drawn attention to the tracks left by native Hawaiians in historic volcanic ash deposits on Kīlauea dating to the late eighteenth century (Moniz Nakamura 2009) using them as analogues with which to interpret the tracks at Laetoli. There are in excess of 1,773 tracks preserved in the Ka'ū Desert on ash previously thought to have been created by the army of the Hawaiian Chief Keōua on his way back from battle in 1790, but now thought to represent a more prolonged period of occupation and transit.

Sticking with the theme of volcanic ash the footprint assemblage from Engare Sero in Tanzania has attracted interest in the research community in recent years (Richmond et al. 2011; Zimmer et al. 2012) but has yet to be documented in full. The site contains at least 350 hominin tracks, formed in a wet volcanic ash that subsequently lithified, in an area of approximately 150 m^2 near the former shore of Lake Natron. The tracks have been variously dated and new dates are currently awaited but may place them as latest Pleistocene, slightly younger than perhaps previously thought. The track assemblage consists of trackways as well as isolated tracks made by several dozen individuals, in a mixed population of children and adults, some of which appear to have been running on the basis of stride lengths. This is an important site not least because of the potential similarity between the volcanic ash and that at the relatively near-by site of Laetoli which might allow for comparison of similar substrates. Further publications on this site are eagerly awaited by the research community. Isolated human tracks have been recorded from the late Pleistocene around the margins of Lake Bogoria in the Kenyan Rift Valley (Scott et al. 2008).

There are two tracksites in South Africa – Nahoon and Langebaan – which both potentially date from the last interglacial around 120 ka BP (Roberts 2008; Jacobs and Roberts 2009). They are both preserved in calcareous aeolianites which makes them distinct within the human track literature. Tracks of various vertebrates were reported from Nahoon Point close to East London in 1964 including three human tracks; originally found as casts on the underside of an overhang, which subsequently collapsed shortly after their discovery (Mountain 1966). Two tracks survived and are now displayed in the East London Museum (Fig. 3.4h). The tracks occur in well-indurated bioclastic-siliclastic sandstone with over 50 % carbonate mostly from comminuted shells. Re-precipitation of the carbonate has led to the lithification of these dune sands which have been dated via OSL to 12.6 ± 8.4 ka The tracks show good anatomical preservation and were probably made by an individual between 1.28 and 1.29 m tall. There is no question that these are human tracks despite the limited number of specimens. The same cannot however be said for the tracks at Langebaan which were found in 1995 by David Roberts on the western shore of the Langebaan Lagoon in Kraal Bay. Two potential human tracks were found along with the eroded remnants of a third and a number of animal tracks. The deposit is again a cemented dune-rock, although the carbonate content is less than at Nahoon (Roberts 2008). The tracks were cast in situ before being removed as a block and transported to the Iziko Museum in Cape Town due to the threat of graffiti from picnic parties (Roberts 2008). The deposit has been dated to around 117 ka, although not with the same precision and consistency as at Nahoon (Robert and Berger 1997). The challenge is that the tracks have relatively poor anatomical form (Fig. 3.4i) and consequently not all authorities are convinced that they are in fact human tracks. However on the basis of the limited trail and the distinct dome-shaped rim structures around the margins of the tracks a human origin remains the most likely interpretation (Roberts 2008). It is interesting within the literature both as an exercise in the correct interpretation of human tracks (see Sect. 2.2) but also because of the importance of the rim structure in making an interpretation. The difference in anatomical preservation between the tracks at Nahoon and those at Langebaan in essentially the same sediment and depositional environment are almost certainly due to the moisture content at the time of imprinting (see Sect. 5.3). The sediment at Langebaan was much drier causing a wider displacement of sediment around the margins of the track, the track-maker also descended diagonally down an asymptotic dune forest which caused enhanced displacement on the downslope side of the tracks as well as some slippage on the plantar surface.

As Lockley et al. (2008) documents there are large number of elusive tracks sites especially in parts of Mexico referred to fleetingly in other publications or via occasional photographs. The specimens and in some case the locations of such sites are not known and they are dated only by association (Rodriguez-de la Rosa et al. 2004). The challenges in tracking down these sites are considerable as illustrated by the tracks at Cuatro Ciénegas discussed above. The site at Oro Grande near Victorville in southern California is notable since the dating seems to be strong with radiocarbon dates being obtained from charcoal rich deposits overlying the tracks (Rector 1979, 1999; Rector et al. 1984). This site and others, including several caves sites, in North America are reviewed by Willey et al. (2005).

The tracks at Willandra Lakes have been document by Webb et al. (2006). The lake system lies within the Murray Basin in south east Australia and consists of 19 interconnected relict lake basins which dried up around 18 ka BP. The human tracks cover an area of around 700 m^2 on an exposed indurated hardpan of laminated calcareous silty clay (150 mm thick), near the shoreline of a small lake basin between the larger Garnpung and Leaghur lakes. They have been dated to between 19 and 23 ka BP based on dates below and above the hardpan which accumulated in a seasonally wet and dry climate. Over 500 human tracks are preserved on the upper surface of the hardpan (Webb 2007). The tracks have good anatomical preservation and well over half of the tracks are contained in up to 23 trackways (Webb 2007) and were made by individuals between 0.86 and 1.98 m tall some of which were running. It is the inferences of running speed that have attracted most attention at this site with some estimates in line with that of world-class athletes today (see Sect. 6.4; McAllister 2011).

3.3 Summary

This brief review of human tracksites illustrates the diversity of depositional environments in which they are typically found. The literature is constantly growing as the recent discovery at Happisburgh illustrates and the potential for future discoveries is likely. In particular the fossil beds of the Koobi Fora Formation in northern Kenya have considerable potential to yield further finds of palaeo-anthropological significance for inter-species comparison. There is a plethora of cave sites which have yet to be documented in detail due to issues of access and method but these sites have potential to yield further insight given the application of modern approaches to data acquisition and analysis. Despite the number of tracksites now documented the comparative study of multiple tracksites, especially those involving different hominin species, requires researchers to examine tracks from different depositional environments and therefore understanding the role of substrate in moderating track topology is crucial something examined further in Chap. 5. It is also important to understand very clearly what can and cannot be inferred from a series of tracks as many of these papers demonstrate. This is something that is explored further in Chap. 6, but is very evident from the literature where a range of claims are made for track-makers with varying levels of reliability.

References

Agnew N, Demas M (1998) Preserving the Laetoli footprints. Sci Am 279:44–55

Aldhouse-Green SHR, Whittle AWR, Allen JRL et al (1992) Prehistoric human footprints from the Severn Estuary at Uskmouth and Magor Pill, Gwent, Wales. Archaeologia Cambrensis 141:14–55

Ambert P, Colomer A, Galant P (2000) Datations mésolithiques des empreintes humaines de l'étage Cathala de la grotte d'Aldène (Cesseras, Hérault). C R Acad Sci Paris, Sciences de la Terre et des planètes/Earth Planet Sci 331:67–74

Aramayo SA (2009) A brief sketch of the Monte Hermoso human footprint site, south coast of Buenos Aires Province, Argentina. Ichnos 16:49–54

Aramayo SA, Manera de Bianco T (2009) Late Quaternary palaeoichnological sites from the Southern Atlantic Coast of Buenos Aires Province, Argentina: mammal, bird and hominid evidence. Ichnos 16:25–32

Aramayo SA, Guitérrez Téllez B, Schillizzi RA (2005) Sedimentologic and paleotologic study of southeast coast of Buenos Aires province, Argentina: a late Pleistocene-Holocene paleoenvironmental reconstruction. J South Am Earth Sci 20:65–71

Ashton N, Lewis SG, De Groote I et al (2014) Hominin footprints from early Pleistocene deposits at Happisburgh, UK. PLoS One 9(2) e88329

Avanzini M, Mietto P, Panarello A et al (2008) The Devil's trails: middle Pleistocene human footprints preserved in a volcanoclastic deposit of Southern Italy. Ichnos 15:179–189

Barnaby W (1975) International News. Nature 254:553

Barriere C, Sahly A (1964) Les empreintes humaines de Lascaux. In: Ripoll E (ed) Miscelanea en homenaje a Abate H. Breuil, vol 1. Diputación Provincial, Barcelona, pp 173–180

Bayon C, Politis G (1996) Estado actual de las investigaciónes en el sitio Monte Hermoso 1 (Provincia de Buenos Aires). Arqueología (Inst. De Ciencias Antropológicas, Facultad de Filosofia y Letras) Universidad de Buenos Aires 6:83–116

Bayón C, Manera T, Politis G et al (2011) Following the tracks of the first South Americans. Evol Educ Outreach 4:205–217
Bayón C, Frontini R, Vecchi R (2012) Middle Holocene settlements on coastal dunes, southwest Buenos Aires Province, Argentina. Quat Int 256:54–61
Beard JB (1997) Lon Odell Memorial Cave, a Dade County treasure. MO Cave Karst Conserv Digest 4:43–48
Begouen CM (1927) Les empreintes de pieds préhistoriques. Institut International d'Anthropologie, III session, Amsterdam, pp 323–327
Behrensmeyer AK (1975) The taphonomy and paleoecology of Plio-Pleistocene vertebrate assemblages east of Lake Rudolf, Kenya. Bull Mus Comp Zool 146:473–578
Behrensmeyer AK, Laporte LF (1981) Footprints of a Pleistocene hominid in northern Kenya. Nature 289:167–169
Bennett MR, Harris JWK, Richmond BG et al (2009a) Early hominin foot morphology based on 1.5 million year old footprints from Ileret, Kenya. Science 323:1197–1201
Bennett MR, Huddart D, Gonzalez S (2009b) Preservation and analysis of three-dimensional footwear evidence in soils: the application of optical laser scanning. In: Dawson L, Miller D, Ritz K (eds) Criminal and environmental soil forensics. Springer, Dordrecht, pp 445–461
Bennett MR, Gonzalez S, Huddart D et al (2010) Probable Neolithic footprints preserved in intertidal peat at Kenfig, South Wales (UK). Proc Geol Assoc 121:66–76
Bennett MR, Morse SA, Falkingham PL (2014a) Tracks made by swimming Hippopotami: an example from Koobi Fora (Turkana Basin, Kenya). Palaeogeog Palaeoclimatol Palaeoecol 409:9–23
Bennett MR, Morse SA, Liutkus-Pierce C et al (2014b) Exceptional preservation of children's footprints from a Holocene footprint site in Namibia. J Afr Earth Sci 97:331–341
Blasi A, Politis G, Bayón C (2013) Palaeoenvironmental reconstruction of La Olla, a Holocene archaeological site in the Pampean coast (Argentina). J Arch Sci 40:1554–1567
Breuil H, Cartailhac É (1907) Une seconde campagne aux cavernes ornées de Niaux (Ariège) et de Gargas (Hautes-Pyrénées). Comptes rendus des séances de l'Académie des Inscriptions et Belles-Lettres 51(4):213–222
Brinton DG (1887) On an ancient human footprint from Nicaragua. Proc Am Philos Soc 24:437–444
Brown RW (1947) Fossil plants and human footprints in Nicaragua. J Paleontol 21:38–40
Brown FH, Haileab B, McDougall I (2006) Sequence of tuffs between the KBS Tuff and the Chari Tuff in the Turkana Basin, Kenya and Ethiopia. J Geol Soc 163:185–204
Bryan AL (1973) New light on ancient Nicaraguan footprints. Archaeology 26:146–147
Casteret N (1948) The footprints of prehistoric man: a vivid new evidence of our ancestors of fifteen thousand years ago. Illus Lond News 110:408–409
Cheong CS, Choi JH, Sohn YK et al (2007) Optical dating of hydromagmatic volcanoes on the southwestern coast of Jeju Island, Korea. Quat Geochronol 2(1):266–271
Chiapella VG (1952) Orsi e nomini preistorici nella grotta della 'Strega', (Genova). Revista del Comune A29:22–29
Cowell RW, Milles A, Roberts G (1993) Prehistoric footprints on Formby Point Beach, Merseyside. In: Middleton R (ed) North West Wetlands survey report, pp 43–48
Crothers GM (1997) Archaeological site inventory form 40RU0090. Submitted to Virginia Department of Historic Resources, Richmond, pp 502–524
Crothers GM, Faulkner CH, Simek JF et al (2002) Woodland cave archaeology in Eastern North America. In: Anderson DG, Mainfort RC Jr (eds) The Woodland Southeast. University of Alabama Press, Tuscaloosa
De Lumley H (1966) Les fouilles de Terra Amata à Nice. Premiers résultats Bulletin du Musée d'Anthropologie Préhistorique de Monaco 13:29–51
De Lumley H (1969) A paleolithic camp at Nice. Sci Am 220:45–50
De Lumley H (1975) Anténéanderthaliens et Néandertaliens du Bassin Mediterranéen occidental et européen. Etud Quat Mém 2 Marseille 1–626
De Lumley H, De Lumley MA, Miskovksy JC et al (1976) Le site de Terra Amata. In: De Lumley H (ed) Sites paléoitiques de la région de Nice et grotte de Grimaldi, Liveret-guide de l'excursion B1, IX" Congrés UISPP, Nice

Deino AL (2011) $^{40}Ar/^{39}Ar$ dating of Laetoli, Tanzania. In: Harrison T (ed) Paleontology and geology of Laetoli: human evolution in context. Springer, Dordrecht, pp 77–97. doi:10.1007/978-90-481-9956-3-4
Di Vito MA, Zanella E, Gurioli L et al (2009) The Afragola settlement near Vesuvius, Italy: the destruction and abandonment of a Bronze Age village revealed by archaeology, volcanology and rock-magnetism. Earth Planet Sci Lett 277(3):408–421
Dillehay T (1989) Monte Verde: a late Pleistocene settlement in Chile: paleoenvironment and site context, vol I. Smithsonian Institutional Press, Washington
Dillehay T (1997) Monte Verde: a late Pleistocene settlement in Chile: the archaeological context, vol II. Smithsonian Institutional Press, Washington
Dingwall HL, Hatala KG, Wunderlich RE et al (2013) Hominin stature, body mass, and walking speed estimates based on 1.5 million-year-old fossil footprints at Ileret, Kenya. J Hum Evol. doi:10.1016/j.jhevol.2013.02.004
Ditchfield P, Harrison T (2011) Sedimentology, lithostratigraphy and depositional history of the Laetoli area. In: Harrison T (ed) Paleontology and geology of Laetoli: human evolution in context. Springer, Dordrecht, pp 47–76. doi:10.1007/978-90-481-9956-3-4
Drake R, Curtis GH (1987) Geochronology of the Laetoli fossil localities. In: Leakey MD, Harris JM (eds) Laetoli: a Pliocene site in northern Tanzania. Clarendon, Oxford
Facorellis Y, Kyparissi-Apostolika N, Maniatis Y (2001) The cave of Theopetra, Kalambaka: radiocarbon evidence for 50,000 years of human presence. Radiocarbon 43(2B):1029–1048
Feibel CS, Brown FH, McDougall I (1989) Stratigraphic context of fossil hominids from the Omo Group deposits: Northern Turkana Basin, Kenya and Ethiopia. Am J Phys Anthropol 78:595–622
Felstead NJ, Gonzalez S, Huddart D (2014) Holocene-aged human footprints from the Cuatrociénegas Basin, NE Mexico. J Archaeol Sci 42:250–259
Garcia MA (1999) La piste de pas humains de la grotte Chauvet à Vallon-Pont d'Arc. Int Newsl on Rock Art 24:1–4
Garcia MA (2001) Les empreintes et les traces humaines et animals. In: Clottes J (ed) La grotte chauvet: l'art des origines. Threshold, Sevil, Paris, pp 34–43
Gonzalez S, Huddart D (2002) Formby Point. In: Huddart D, Glasser NF (eds) Quaternary of northern England. JNCC, Peterborough, pp 569–582
Gonzalez AH, Lockley MG, Rojas C et al (2007) Notes on the re-discovery of a 'lost' hominid footprint site from the Cuatro Cienegas basin (Coahuila), Mexico. N M Mus Nat Hist Sci Bull 42:11–15
Gonzalez AH, Lockley MG, Rojas CS et al (2009) Human tracks from quaternary tufa deposits, Cuatro Ciénegas, Coahuila, Mexico. Ichnos 16:12–24
Haberland W, Grebe WH (1957) Prehistoric footprints from El Salvador. Am Antiq 22(3):282–285
Harrington SPM (1999) Human footprints at Chauvet Cave. Archaeology 52(5):18
Harris JM (ed) (1983) Koobi Fora research project volume 2: the fossil ungulates proboscidea, perissodactyla and suideae. Clarendon, Oxford
Harris JM (ed) (1991) Koobi Fora research project volume 3: stratigraphy artiodactyls and palaoenvironments. Clarendon, Oxford
Hay RL (1987) Geology of the Laetoli area. In: Leakey MD, Harris JM (eds) Laetoli: a Pleistocene site in northern Tanzania. Clarendon, Oxford
Hooijer DA (1987) Hipparions of the Laetolil beds, Tanzania. In: Leakey MD, Harris JM (eds) Laetoli: a Pliocene site in northern Tanzania. Clarendon, Oxford
Huddart D, Gonzalez S, Roberts G (1999a) The archaeological record and mid-Holocene marginal coastal paleoenvironments around Liverpool Bay. J Quat Sci 7:563–574
Huddart D, Roberts G, Gonzalez S (1999b) Holocene human and animal footprints and their relationships with coastal environmental change, Formby Point, NW England. Quat Int 55:29–41
Isaac GLI, Behrensmeyer AK (1997) Geological context and palaeoenvironments. In: Isaac GLI (ed) Koobi Fora research project. Plio-Pleistocene archaeology, vol 5. Clarendon, Oxford, pp 12–70

Jacobs Z, Roberts DL (2009) Last interglacial age for aeolian and marine deposits and the Nahoon fossil human footprints, southeast coast of South Africa. Quat Geochronol 4(2):160–169
Jaggar TJ (1921) Fossil human footprints in Kau Desert. Hawaii Volcano Obs Mon Bull 9:114–118
Jaggar TA (1934) Fossil human footprints in the Kau Desert. Paradise Pac 46(5):5
Kim JY, Kim KS, Lockley M, Matthews N (2008) Hominid ichnotaxonomy: an exploration of a neglected discipline. Ichnos 15:126–139
Kim KS, Kim JY, Kim SH et al (2009) Preliminary report on hominid and other vertebrate footprints from the Late Quaternary strata of Jeju Island, Korea. Ichnos 16:1–11
Kim CB, Kim JY, Kim KS et al (2010) New age constraints for hominid footprints found on Jeju Island, South Korea. J Archaeol Sci 37(12):3338–3343
Kinahan J (1996) Human and domestic animal tracks in an archaeological lagoon deposit on the coast of Namibia. South Afr Archaeol Bull 51:94–98
Kinahan J (2001) Pastoral nomads of the Central Namib Desert: the people history forgot, 2nd edn. Namibia Archaeological Trust, Windhoek
Kinahan J (2013) The use of skeletal and complementary evidence to estimate human stature and identify the presence of women in the recent archaeological record of the Namib Desert. South Afr Archaeol Bull 68(197):72–78
Kinahan J, Pallett J, Vogel JC et al (1991) The occurrence and dating of elephant tracks in the silt deposits of the lower! Kuiseb River, Namibia. Cimbebasia 13:37–43
Laporte LF, Behrensmeyer AK (1980) Tracks and substrate reworking by terrestrial vertebrates in Quaternary sediments of Kenya. J Sedimentol Res 50:1337–1346
Leakey MD (1981) Tracks and tools. Philos Trans R Soc Lond B Biol Sci 292(1057):95–102
Leakey MD, Harris JM (1987) Laetoli: a Pliocene site in northern Tanzania. Clarendon, Oxford
Leakey MD, Hay RL (1979) Pliocene footprints in the Laetoli beds at Laetoli, northern Tanzania. Nature 278:317
Leakey MG, Leakey REF (eds) (1978) Koobi Fora research project volume 1: the fossil hominids and an introduction to their context, 1968–1974. Clarendon, Oxford
Lepre CJ, Quinn RL, Joordens JC, Swisher CC, Feibel CS (2007) Plio-Pleistocene facies environments from the KBS Member, Koobi Fora formation: implications for climate controls on the development of lake-margin hominin habitats in the northeast Turkana Basin (northwest Kenya). J Hum Evol 53:504–514
Lockley MG, Kim JY, Roberts G (2007) The Ichnos project: a re-evaluation of the hominid track record. N M Mus Nat Hist Sci Bull 42:79–89
Lockley M, Roberts G, Kim JY (2008) In the footprints of our ancestors: an overview of the hominid track record. Ichnos 15:106–125
Lockley MG, Vasquez RG, Espinoza E et al (2009) America's most famous human footprints: history, context and first description of mid-Holocene tracks from the shores of Lake Managua, Nicaragua. Ichnos 16:55–69
Mafart B (2006) Anthropological study of Holocene human footprints of the sebkra El Azrag (Islamic Republic of Mauritania). Anthropologie 110(5):766–787
Marcos EL (2001) El descubrimiento de las Huellas Prehistoricas. Ojo Guarena. Cubía, Boletín del Grupo Espeliológico Edelweiss June 3:34–35
Mastrolorenzo G, Petrone P, Pappalardo L et al (2006) The Avellino 3780-yr-B.P. catastrophe as a worst case scenario for a future eruption at Vesuvius. Proc Natl Acad Sci U S A 103:4366–4370
McAllister PJ (2011) The evolution of the inadequate modern male. Aust Sci May:19–21
McHenry LJ (2011) Geochemistry and mineralogy of Laetoli area tuffs: lower Laetolil through Naibadad beds. In: Harrison T (ed) Paleontology and geology of Laetoli: human evolution in context. Springer, Dordrecht, pp 121–142. doi:10.1007/978-90-481-9956-3-4
Meldrum DJ (2000) Footprints in the Ka'u Desert, Hawaii. Am J Phys Anthropoll Suppl 30:226–227
Meldrum DJ (2002) Midfoot flexibility and the evolution of bipedalism. Am J Phys Anthropoll Suppl 34:111–112

Meldrum DJ (2004a) Fossilized Hawaiian footprints compared with Laetoli hominid footprints. In: Meldrum DJ, Hilton CE (eds) From biped to strider: the evolution of modern human walking, running, and resource transport. Kluwer-Plenum, New York
Meldrum DJ (2004b) Midfoot flexibility, fossil footprints, and Sasquatch steps: new perspectives on the evolution of bipedalism. J Sci Explor 18:65–79
Meldrum DJ (2007) Renewed perspective on the Laetoli trackways: the earliest hominid footprints. In: Lucas SG, Spielmann JA, Lockley MG (eds) Cenozoic vertebrate tracks and traces. N M Mus Nat Hist Sci Bull 42:233–239
Meldrum DJ, Lockley MG, Lucas SG et al (2011) Ichnotaxonomy of the Laetoli trackways: the earliest hominin footprints. J Afr Earth Sci 60:1–12
Meltzer DJ, Grayson DK, Ardilla G et al (1997) On the Pleistocene antiquity of Monte Verde, southern Chile. Am Antiq 62(4):659–663
Mietto P, Avanzini M, Rolandi G (2003) Palaeontology: human footprints in Pleistocene volcanic ash. Nature 422:133
Miskovski JC (1967) Les paleosols Mindel-Riss et Riss-Wurm de Terra Amata. Etude sédimentologique, Académie Science Paris 264:2361–2363
Molleson TI, Oakley KP, Vogel JC (1972) The antiquity of the human footprints of Tanna della Basura. J Hum Evol 1:467–471
Moniz Nakamura MJJ (2009) Hominid footprints in recent volcanic ash: new interpretations from Hawaii volcanoes National Park. Ichnos 16:118–123
Morse SA, Bennett MR, Liutkus-Pierce C et al (2013) Holocene footprints in Namibia: the influence of substrate on footprint variability. Am J Phys Anthropol 151:265–279
Mountain ED (1966) Footprints in calcareous sandstone at Nahoon Point. South Afr J Sci 62:103–111
Musiba CM, Mabula A, Selvaggio M et al (2008) Pliocene animal trackways at Laetoli: research and conservation potentials. Ichnos 15:166–178
Noval Fonseca MA (2007) Cuevas tempranas (Posada, Llanes): Fragil Rastro del Pasado. Excavaciones arqueologicos en Asturias 1999–2002, Gobierno del Principado de Asturias. Servicio de Publicaciones, Oviedo, pp 207–213
Onac BP, Viehmann I, Lundberg J et al (2005) U–Th ages constraining the Neanderthal footprint at Vârtop Cave, Romania. Quat Sci Rev 24(10):1151–1157
Ozansoy F (1969) Pleistocene fossil human footprints in Turkey. Bull Mineral Res Explor Inst Turkey 72:146–150
Pales L (1954) Les empreintes de pieds humains de la “Tana della Basura” (Toriano). Revista di Studi Liguri 20:5–12
Pales L (1960) Les empreintes de pieds humains de la “Grotta della Basura” (Toriano). Revue des études ligures 26:25–90
Pales L (1976) Les empreintes de pieds Humains dans les cavernes. Archives de l’ Institut de Paléontologie Humaine 36:1–166
Parfitt SA, Barendregt RW, Breda M et al (2005) The earliest record of human activity in northern Europe. Nature 438:1008–1012
Parfitt SA, Ashton NM, Lewis SG et al (2010) Early Pleistocene human occupation at the edge of the boreal zone in northwest Europe. Nature 466:229–233
Pobiner BL, Rogers MJ, Monahan CM et al (2008) New evidence for hominin carcass processing strategies at 1.5 Ma, Koobi Fora, Kenya. J Hum Evol 55(1):103–130
Rector CH (1979) 5,000-year-old footprints on the Mojave River, California, USA. Antiquity 54:149–150
Rector CH (1999) Human and animal trackway at Oro Grande. San Bernardino County Mus Assoc Q 46:53–55
Rector C, Swenson JD, Wilke PJ (1984) Archaeological studies at Oro Grande, Mojave Desert. Calif J Calif Gt Basin Anthropol 6(2):280–282
Renfrew C, Morley I (eds) (2009) Becoming human: innovation in prehistoric material and spiritual culture. Cambridge University Press, Cambridge
Richmond BG, Hatala KG, Harcourt-Smith WEH et al (2011) Early modern human footprint assemblage from Engare Sero, Tanzania. Paleoanthropology 2011:A31

Robbins LM, Wilson RC, Watson PJ (1981) Paleontology and archaeology of Jaguar Cave, Tennessee. Proc Eighth Intern Congr Speleol 1:377–380
Roberts D (2008) Last interglacial hominid and associated vertebrate fossil trackways in coastal eolianites, South Africa. Ichnos 15(3/4):190–207
Roberts DL, Berger LR (1997) Last interglacial (c. 117 kyr) human footprints from South Africa. S Afr J Sci 93:349–350
Roberts G, Gonzalez S, Huddart D (1996) Inter-tidal Holocene footprints and their archaeological significance. Antiquity 70:647–651
Rodriguez-de la Rosa RA, Aguillon-Martinez MC, Lopez-Espinosa J (2004) The fossil record of vertebrate tracks in Mexico. Ichnos 11:27–37
Scaillet S, Vita-Scaillet G, Guillou H (2008) Oldest human footprints dated by Ar/Ar. Earth Planet Sci Lett 275(3):320–325
Schmincke H-U, Kutterolf S, Perez W, Rausch J, Freundt A, Strauch W (2009) Walking through volcanic mud: the 2,100 year old Acahualinca footprints (Nicaragua). Bull Volcanol 71:479–493
Schmincke HU, Rausch J, Kutterolf S, Freundt A (2010) Walking through volcanic mud: the 2,100 year-old Acahualinca footprints (Nicaragua) II: the Acahualinca people, environmental conditions and motivation. Int J Earth Sci 99:279–292
Scott JJ, Renaut RW, Owen RB (2008) Preservation and paleoenvironmental significance of a footprinted surface on the Sandai Plain, Lake Bogoria, Kenya Rift Valley. Ichnos 15:208–231
Sneed JM (1984) The Sequoyah Caverns story. Nat Speleol Soc News 42:178–181
Suwa G (1984) Could Australopithecus afarensis have made the hominid tracks at Laetoli? J Phys Anthropol 963:224–225
Tuttle R (2008) Footprint clues in hominid evolution and forensics: lessons and limitations. Ichnos 15:158–165
Tuttle R, Webb D, Weidl E et al (1990) Further progress on the Laetoli trails. J Archaeol Sci 17:347–362
Valladas H, Tisnerat-LaBorde N, Cachier H et al (2005) Bilan de datations carbone 14 effectuées sur des charbons de bois de la grottee Chauvet. Bull Soc Préhistorique Française 102:109–113
Vallois HV (1927) Etude des empreintes de pieds humains de Tuc d'Audoubert, de Cabrerets, de Ganties. Institut International d'Anthropologie, III session, Amsterdam, 328–338
Vallois HV (1931) Empreintes de pieds humains des Grottes préhistorique du midi de la France. Paleobiologica 4:79–98
Watson PJ (1969) The prehistory of Salts Cave, Kentucky. Illinois State Museum Reports of Investigations 16
Watson PJ (1974) Archaeology of the Mammoth Cave Area. Academic, New York
Watson PJ (1982) Notes on a visit to Fisher Ridge Cave. Unpublished field notes dated June 27, 1982
Watson PJ (1986) Prehistoric cavers in the Eastern Woodlands. In: Faulkner CH (ed) The prehistoric native American Art of Mud Glyph Cave. University of Tennessee Press, Knoxville
Watson PJ, Kennedy MC, Willey P et al (2005) Prehistoric footprints in Jaguar Cave, Tennessee. J Field Archaeol 30:25–43
Webb S (2007) Further research of the Willandra Lakes fossil footprint site, south eastern Australia. J Hum Evol 52:711–715
Webb S, Cupper ML, Robins R (2006) Pleistocene human footprints from the Willandra Lakes, south eastern Australia. J Hum Evol 50:405–413
Westaway R, Pringle M, Yurtmen S et al (2004) Pliocene and Quaternary regional uplift in Western Turkey: the Gediz river terrace staircase and the volcanism at Kula. Tectonophysics 391:21–169
Westaway R, Guillou H, Yurtmen S et al (2006) Late Cenozoic uplift of western Turkey: improved dating of the Kula Quaternary volcanic filed and numerical modeling of the Gediz River terrace staircase. Global Plan Change 51:131–171

White TD, Suwa G (1987) Hominid footprints at Laetoli: Facts and interpretations. Am J Phys Anthropol 72:485–514

Willey P, Stolen J, Crothers G et al (2005) Preservation of prehistoric footprints in Jaguar Cave, Tennessee. J Cave Karst Stud 67(1):61–68

Willey P, Watson PJ, Crothers G et al (2009) Holocene human footprints in North America. Ichnos 16:70–75

Williams H (1952) Geologic observations on the ancient human footprints near Managua, Nicaragua. Carnegie Inst Wash Publ (Contrib Am Anthro Hist 11) 596:1–32

Wynn JG (2004) Influence of Plio-Pleistocene aridification on human evolution: Evidence from paleosols of the Turkana Basin, Kenya. Am J Phys Anthropol 123:106–118

Zhang DD, Li SH (2002) Optical dating of Tibetan human hand- and footprints: an implication for the palaeoenvironment of the last glaciation of the Tibetan Plateau. Geophys Res Lett 29. doi: 10.1029/2001GL013749

Zimmer B, Liutkus C, Carmichael S et al (2012) A snapshot in time determining the age, environment and social structures of Early Homo sapiens using trace fossils in volcanic rocks at the Engare Sero footprint site, Lake Natron, Tanzania. https://gsa.confex.com/gsa/2012CD/finalprogram/abstract_201451.htm

Chapter 4
Geoconservation of Human Tracks

Abstract In the previous chapter we have seen how there is a wide diversity of human tracksites each with a different depositional history and mechanism of track preservation. The challenge for the geoarchaeologist is not only to document such sites but also to advise with respect to their long-term conservation. At many human tracksites this is challenging due to the nature of the soft, erodible substrate and the geomorphological environments in which the tracks are now exposed. In this chapter we explore some of these challenges and suggest some solutions.

4.1 Geoconservation

Both with respect to fundamental founding research as well as practical action the idea that our geological heritage (geoheritage) should be conserved alongside and in equal measure to our natural and cultural heritage has developed rapidly in the last 20 years through the work of numerous dedicated geoscientists (Gray 2002; Dowling 2011; Henriques et al. 2011; Hose 2012; Matthews 2013). The seminal book by Gray (2002) brought much of the history and emerging ideas together in one place for the first time and the advent in 2009 of the journal *Geoheritage* published by Springer was another academic landmark in this journey. Geoheritage conservation still remains a poor relation in some respects to the more easily understood issues of wildlife, biodiversity and archaeological heritage conservation (e.g., Wilson 1994; Doyle and Bennett 1998; Page 1998; Reis and Henriques 2009) and the case continues to be made for the need to see all types holistically (Matthews 2013). The point is particularly well made with respect to human tracksites which represent a combination of geology and archaeology/anthropology.

Geoheritage conservation is for the most part still based on the concept of a 'boundary' within which heritage content is identified, valued and bounded geographically in some way to be afforded in various country specific ways legal, planning, and/or development protection (Doyle and Bennett 1998). The way in which this is implemented may vary from region to region and country to country as does the value base that is used to define the singular, or multiple, content of a site and its geographical scale. In Europe this is tied with the emergence of the 'geotope' namely a distinct part of the geosphere of outstanding geological/geomorphological

M.R. Bennett, S.A. Morse, *Human Footprints: Fossilised Locomotion?*,
DOI 10.1007/978-3-319-08572-2_4

interest to be afforded recognition and/or protection in some way (Strüm 1994; Gray 2002). UNESCO focused on creating a list of Global Geosites – those sites of first-class importance to geology – and more recently on the concept of the 'geopark' (Wimbledon 1996; Gray 2002). The 'geopark' is again a boundary defined concept but generally on a much larger scale. It encompasses one or more site(s) of geological importance, but crucially in association with archaeological, ecological or cultural values and is driven by a management plan that focuses on the economic needs of the local population through growth of geotourism, whilst attempting to safeguard the heritage and the landscape in which that population lives (Eder and Patzak 2004; UNESCO 2006; Hose 2012; Farsani et al. 2014). The concept has become popular in Europe with parks located in countries as far apart as Greece and Ireland and there is now an established European Geopark Network (Marty et al. 2004). The degree to which this aim of driving rural economic development is achieved, especially without compromise to the conservation agenda, is dependent on the case in question (e.g., Hose 2012; Kiernan 2013).

The conservation of vertebrate tracksites, mainly those made by dinosaurs, has been achieved via a range of different designations across the World, including geoparks in Europe (e.g., Lockley and Meyer 1997; Marty et al. 2004; Santos et al. 2008) and a wide range of different conservation strategies have been adopted in the management and protection of these sites (Agnew et al. 1989; Marty et al. 2004). They are perceived by some as 'an easy sell' in terms of conservation having both popular appeal and educational value (Reis and Henriques 2009). Human tracksites are a subset of these sites, but in practice share few similarities, due to the fragile nature of such sites associated as they are for the most part with soft and erodible sediment (Bennett et al. 2013). Our aim here is therefore to explore the specific issues that pertain to human tracksites and to steer clear of wider discussions around geoheritage conservation.

4.2 Placing Value on Human Tracksites

Perhaps the most fundamental question to ask is why should we try to conserve a human tracksite at all? As scientists we tend to see 'value' in terms of the scientific contribution a site may have to our understanding of a particular problem. Clearly in this context some sites have greater geological, archaeological or palaeoanthropological significance than others. However this is only one 'value' measure that could be applied (Gray 2002). For example, a site may be of spiritual importance to a particular community, be of economic value to its tourism industry (Hose 2012), or simply be a source of local civic pride. At the heart of this debate are ethical and philosophical questions (Beckerman and Pasek 2001) around the intrinsic and extrinsic value of something; does something have value for being simply what it is, or does value come from its utility in human service? While one can see how intrinsic value may apply to a living thing, does it apply to something that isn't like a human track? These are not easy questions and ultimately a matter of opinion and perspective as well as cultural norms (Gray 2002). Within the UK at least four main value systems have been recognised with respect to geoconservation (Bennett and Doyle 1997; Doyle and Bennett 1998), namely: intrinsic value; cultural and aesthetic value; economic value; and research and educational value. To this classification Gray (2002) added the functional value in the context of Earth resources and processes (cf. Wimbledon 2006). While it is relatively easy, although sometimes controversial, to quantify economic value, for example via some form of cost-benefit analysis, it is much harder to assign a monetary value to the aesthetics or cultural importance of particular geological phenomena or deposits (Doyle and Bennett 1998). For example, 'beauty lies in the eyes of the beholder' and the aesthetic appeal of a particular landscape or artefact is very much influenced by cultural and social history as well as fashion. Again there are no easy answers here, but awareness of the complexity of the issues is important. Reis and Henriques (2009) take an interesting approach (Fig. 4.1); they recognise two value classes one which they call 'Relevance Grade' and a second called 'Abstract Perceptiveness'. Relevance Grade refers to the geological content of a site and its scientific importance defined by consensus within the scientific community. They argued that while this may make sense to geologists it may not be widely understood or valued by society and this is where their Abstract Perspectives comes into play since this depends on the public understanding of that content. Within this matrix they define a number spaces through terms such as: Indicial, content which on a local scale demonstrates clear linkage between geological process and product; Iconographic, content which inspires personal imagination such as a trackway; Symbolic, geological content which is associated with locations where people naturally gather for example like the White Cliffs of Dover in the UK; Documental, content that is well documented and critical to the geological understanding of a region or time period in earth history; Scenic, content associated or linked to a high level of recreational function; and Conceptual, content which demonstrates global phenomena or underpins key theoretical principles in geology that transcend a particular site. Reis and Henriques (2009) argue that the challenge is to steer a path between 'scientific' and 'social' influences combing the two. It is simply one of many ways of ascribing value to geological phenomena such as human tracks (e.g., Mampel et al. 2009; Bruschi et al. 2011; Fassoulas et al. 2012), but crucially it recognises the importance of the interplay between scientific and social/cultural value systems. In the context of human tracksites, the point we are trying to make is that as scientists it is easy to slip into considering 'value' simply in terms of a sites research contribution when in fact the value or importance of a site may be much greater when the views of all the stakeholders are considered and viewed equally (Gray 2002).

In reviewing the conservation options for the Laetoli tracksite Demas and Agnew (2006) provided a framework in which such discussion could proceed, although many of the issues overlap.

- Values. Demas and Agnew (2006) discuss this in the context of the research contribution a site might and can make both at the time of discovery, currently and its future potential although they recognise as we have above that other

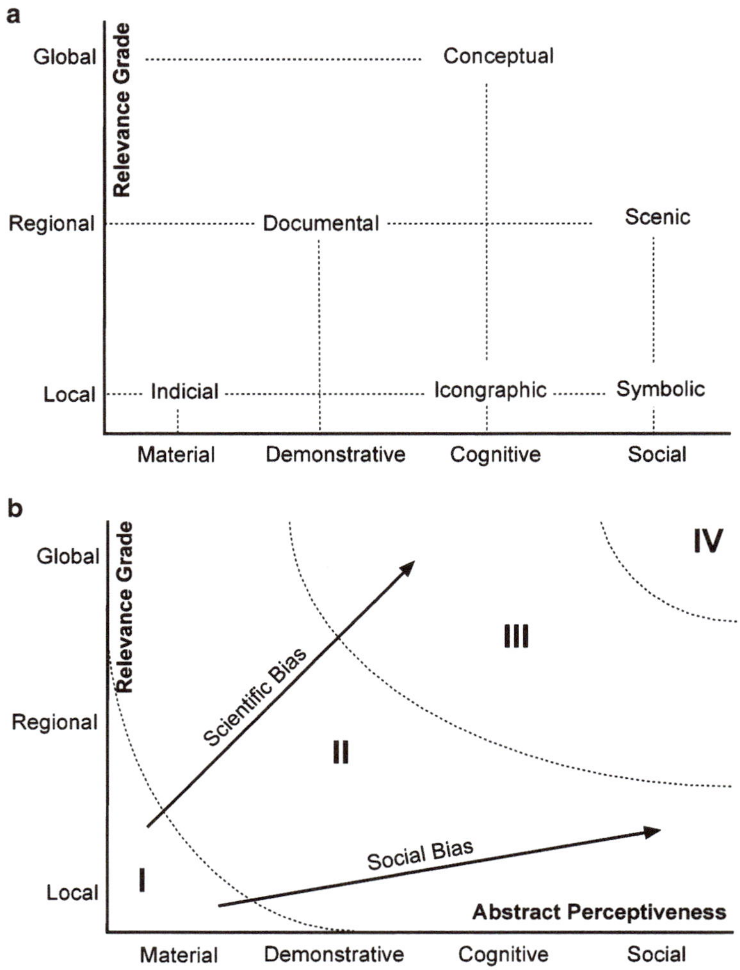

Fig. 4.1 An example of an integrated evaluation system for geoheritage using both scientific and cultural/social value systems. (**a**) The basic matrix with the Relevance Grade and the Abstract Perceptiveness defined. (**b**) Defined areas of value I–IV based on a combination of both axis, the potential for scientific and social biases is shown by the two *arrows* (Modified from Reis and Henriques (2009))

'value' systems may be equally important. For example, the Devil's Tracks at Roccamonfina (see Sect. 3.1.5) are of immense civic pride to the local community and have a rich connection to local folklore (Mayor and Sarjeant 2001). Equally the Namibian footprints at Walvis Bay support a number of local businesses centred on heritage tourism and are fiercely protected by some of the principals involved. There is also the symbolic and spiritual value of human tracksites as stated by Demas and Agnew (2006, p. 67) 'footprints offer a unifying and potent symbol of our species and our beginnings'. The Sefton Coast tracks have inspired writers and are linked to modern almost spiritual discussions of the landscape wonderfully invoked by Macfarlane (2012) in his book the *Old Ways*. It is easy for scientists informing decision-makers to lose sight of these wider judgements of value as argued above.

- Benefits. The benefits to the scientific community and the quest to understand human origins and evolution are considerable and taken as read here but of course link to the 'value' base that is being applied and as we have argued other 'value' systems may apply. Potential financial benefits to a local community stem from tourism that a site may attract (Hose 2012). The challenge here is that many of the benefits are hard to quantify in terms of monetary value and therefore often difficult to set against any capital investment needed to conserve the site or to provide the tourist infrastructure necessary for them to generate revenue for an area or region.
- Stakeholders. The stakeholders are multiple but include the land owner and any associated public bodies both locally and nationally that have an interest in the site in the context of planning or heritage conservation. The original excavators/researchers, as well the current excavators and permit holders all have a vested interest as well as the scientific community at large who may want site access or have an interest in the emerging findings. However it is the local and community interests that are sometimes neglected here since they are often difficult to engage with being loose collections of enthusiasts and amateurs who rarely speak with one voice. The more complex the value-base being applied the wider and more varied the list of stakeholders needs to be and the more likely that both conscious and subconscious prejudices may emerge that do not allow all views to be treated with the equality they perhaps deserve.

In practice value and benefits are informed by the stakeholders involved and crucially those that perform or broker the evaluation as well as the resources available to devote to the conservation of a particular tracksite. The conditions that pertain to a particular site and the threats that it faces are also critical in determining the correct conservation option.

4.3 Conservation Risks: Threats and Challenges

Human tracksites are preserved in a range of different depositional environments which now outcrop, and are exposed, in a variety of geomorphological settings (Lockley et al. 2008; Table 3.1) and as such are exposed to different levels of conservation risk. For example, the footprints of the Sefton Coast in northern England (Roberts et al. 1996) were originally formed in coastal lagoons and on mudflats behind a coastal barrier, but are now exposed on the current foreshore when storm events draw-down protective beaches exposing the underlying ichnologically-rich silt beds (Fig. 4.2a, b). The tracks at Monte Hermoso (Argentina; Fig. 4.2c) were

Fig. 4.2 Vulnerable human track sites. (**a**) General view of eroding Sefton Coast (UK) with exposed human tracks. (**b**) Track way on the Sefton Coast. (**c**) Human tracks at Monte Hermoso, these tracks were formed inland, but are now exposed beneath thin beach sediment on the Atlantic coast of Argentina

formed inland within ephemeral fluvial systems but are now exposed beneath a thin layer of beach sand at the coast. In Namibia, the human tracks at Walvis Bay (see Sect. 3.2.3) are exposed on terrace surfaces formed of fluvial over-bank flood deposits which were desiccated, imprinted and subsequently buried by mobile sand dunes and are exposed periodically as dunes migrate over the surface (Morse et al. 2013). These prints are quickly eroded and deflated when the salt hardened silt is disturbed either naturally or increasingly by tourists and by recreational vehicles exploring the dune fields. At these sites the continued operation of dune migration is important to expose new surfaces. More generally the continued operation of geomorphological processes is important at other sites too. For example, at Nahoon in South Africa a trail of human prints is preserved in aeolianites dated to 124±4 ka BP and are potentially amongst the oldest examples of *Homo sapiens* prints in Africa (Roberts 2008). The Nahoon prints were first observed in 1964 within an overhang which then collapsed with two prints being rescued and transported to the local East

London Museum (Roberts 2008). The original site, on the sea cliffs at Nahoon Point, is maintained as a tourist attraction with a wonderful footprint-shaped visitors centre alongside an excavation preserve. One could argue at a site such as Nahoon, that continued coastal erosion and cliff collapse has the potential to reveal new prints in time, despite the risk that some may be lost during that process and there is a balance to be struck here of course. The potentially important human tracks exposed by erosion at Happisburgh on the Norfolk Coast (Ashton et al. 2014; see Sect. 3.1.4) were exposed and lost through erosion before they could be properly documented, but continued erosion will hopefully in time reveal new tracks. The conservation strategies relevant to each of these sites is potentially different but allowing natural process to continue – dune migration and coastal erosion for example – is potentially important since even though tracks may be lost new surfaces and tracks may be revealed.

This sits in contrast to the approach adopted at Laetoli in northern Tanzania. The prints were first excavated in 1978–1979 and documented by photogrammetry and extensive casting of the surface (Day and Wickens 1980; Leakey and Harris 1987). The site was originally re-buried with backfill which was unfortunately rich in acacia seeds, and as these seeds germinated and grew the site became threatened by damage from roots leading to the site being re-excavated in the 1990s (Agnew and Demas 1998, 2004; Demas and Agnew 2006; Musiba et al. 2008). Agnew and Demas (1998) documented the decision-making process which led to the re-burial of the site in a controlled manner involving the use of a clean backfill and a range of geo-membranes to act as a root barrier. The solution while effective (Agnew and Demas 2004) remains an area of continual tension both for the scientific community who are denied access to the tracks and because of the inability both locally and nationally to derive tourist and cultural value from the site. A range of alternative plans have been mooted in recent years including the complete excavation and removal to some form of either local or national museum (Musiba et al. 2008). The strategy here is clear, to preserve at all costs a finite resource, something which is aided by the fact that although fragile, the substrate is partially lithified.

Bennett et al. (2013) argue that there is a spectrum of human tracksites from those of relatively low scientific value, often with high print numbers, which are located in sites that are threatened continually by natural processes but are in turn dependent to some degree on natural processes for exposure (e.g., Sefton Coast or Namibia), via those at Nahoon or Langebaan which are of greater scientific significance and are more limited in extent, to those at the other extreme such as Laetoli which are of considerable scientific importance, limited in number, but are preserved in a comparatively firm substrate and are consequently threatened less by natural processes (Fig. 4.3). The Ileret footprints of northern Kenya (Bennett et al. 2009; see Sect. 3.1.2) sit uncomfortably within this spectrum, being arguably of considerable scientific importance, modest in number but preserved in what is highly erodible and unlithified sediment. The key point here is that the Ileret site is located in an eroding bluff of horizontal beds which have been excavated at various levels (Bennett et al. 2009; Dingwall et al. 2013; Fig. 4.4a). The slightly younger prints found 45 km to the south close to Koobi Fora (Behrensmeyer and Laporte 1981; see

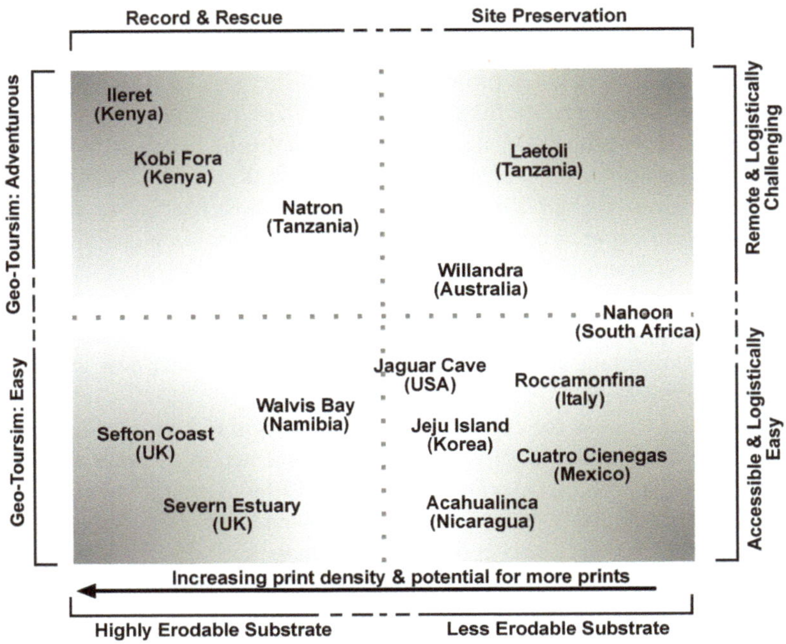

Fig. 4.3 Matrix of variables relevant to the conservation of hominin/human footprint sites with particular emphasis on soft-sediment sites following the Bennett et al. (2013). The horizontal continuum at the top is between strategies based on 'record and rescue' versus those based on site preservation either via burial such as Laetoli or via some form of conserved display as is the case at Acahualinca in Nicaragua

Sect. 3.1.3) are preserved in similar sediments but in this case they dip at 15–18° to the west into one of the valley sides and occur in the valley floor close to a dry river channel (Fig. 4.4b). The key conservation threats at both of these Kenyan sites can be summarised as: (1) sediment weathering and gravity driven slope failure; (2) lateral fluvial erosion and/or rain induced gullying during wet seasons; (3) bio-erosion due to roots, animal burrows and the passage of grazing livestock; (4) unlawful excavation; (5) inappropriate exploitation by indigenous populations, for example removing valuable assets such as plastic sheeting used for preservation between excavation seasons, or accidental, curiosity driven damage; (6) damage during repeated re-excavation during successive field seasons; and (7) break-up of the sediment surfaces due to changes in sediment moisture content (causing swelling or desiccation), thermal expansion/contraction and vertical unloading all of which can be caused by changes to overburden volume, surface run-off and hydrogeology during excavation of benches and introduction of plastic sheeting and other impermeable membranes by excavators. Of these the most important are probably changes to the sediment moisture content and natural erosional processes in semi-arid environments. These sites provide, as all soft-sediment human tracksites do, some real conservation challenge.

Fig. 4.4 Threats to the conservation of soft-sediment sites. (**a**) The Ileret hominin track site in northern Kenya. The tracks are exposed at multiple levels within this eroding bluff, see Sect. 3.1.2 for further details. (**b**) Koobi Fora hominin track site in northern Kenya. The palaeosurface dips into the slope and the site is located at the base of a dry river valley providing both a challenge for the excavator and for site conservation. It is worth noting that part of this surface survived for over 30 years buried beneath backfill and with a layer of plastic/canvas

4.4 Conservation Options

Bennett et al. (2013) recognised three broad options for the conservation of human tracksites: (1) on-site conservation; (2) off-site conservation; and (3) do-nothing. Options for on-site preservation include a range of possibilities from exposed display to some form of shelter. A wide range of options have been adopted to conserve dinosaur track sites as illustrated in Figs. 4.5 and 4.6 however not all these solutions

Fig. 4.5 Protection measures for dinosaur footprint sites (tracksites); all these photographs are reproduced with kind permission of Daniel Marty. (**a**) Cretaceous Hwasun theropod, ornithopod and sauropod tracksite (Huh et al. 2006; Kim and Huh 2010; Lockley et al. 2012) in South Korea (Seoyu-ri, Hwasun County, Jeollanamdo Province, National Monument 487). There are two main track-bearing surfaces, one of which is surrounded by a trail. Both palaeosurfaces are inclined and exposed to erosion. (**b**) Early Cretaceous Gajin sauropod, theropod, and bird tracksite (Falk et al. 2010; Kim and Lockley 2012; Kim et al. 2012) in South Korea (Gajin-ri, Jinseong-myeon, Jinju city, South Gyeongsang Province, Natural Monument 395). This tracksite was discovered 1997 during the construction of the Gyeongnam Institute of Science Education and it was protected by directly integrating two palaeosurfaces as Fossil Heritage Halls I and II into the new building. (**c**) Late Cretaceous Unhangri dinosaur, pterosaur, and bird tracksite (Lee and Huh 2002; Hwang et al. 2002, 2008) in South Korea (Haenam-gun County, South Jeolla Province, Natural Monument 394). This picture shows a palaeosurface with pterosaur tracks that was protected by the construction of a

Fig. 4.6 Protection measures for dinosaur footprint sites (tracksites); all these photographs are reproduced with kind permission of Daniel Marty. (**a**) Early Cretaceous Las Cerradicas sauropod and ornithopod tracksite (Pérez-Lorente et al. 1997; Castanera et al. 2011) in Spain (Galve, Teruel Province) was protected by the construction of a metallic building as part of the Dinosaur Ichnites of the Iberian Peninsula Project (IDPI). This building is, however, not entirely closed and the site is still exposed to a certain degree of weathering. (**b**) Late Cretaceous Unhangri dinosaur, pterosaur, and bird tracksite (Lee and Huh 2002; Hwang et al. 2002, 2008) in South Korea (Haenam-gun County, South Jeolla Province, Natural Monument 394). This picture shows the building that protects the palaeosurface with pterosaur tracks shown in Fig. 4.5c

Fig. 4.5 (continued) closed building. This site is part of the Uhangri Dinosaur Museum. The outside of this site is shown in Fig. 4.6b. (**d**) Late Jurassic Barkhausen sauropod and theropod tracksite (Kaever and Lapparent 1974; Diedrich 2010) in N Germany (Wiehengebirge, near Osnabrück). The strongly inclined track-bearing palaeosurface was covered with a roof but is still exposed to weathering, notably by water income from above (along the bedding plane). (**e**) Early Cretaceous Las Cerradicas sauropod and ornithopod tracksite (Pérez-Lorente et al. 1997; Castanera et al. 2011) in Spain (Galve, Teruel Province). The inclined palaeosurface was protected by the construction of a metallic building including a visitor platform (*right* in the picture) as part of the Dinosaur Ichnites of the Iberian Peninsula Project (IDPI). This building is, however, not entirely closed and the site is still exposed to a certain degree of weathering. (**f**) Early Cretaceous sauropod and ornithopod (Lockley et al. 2004) Münchehagen tracksite in N Germany (near Hannover). The horizontal palaeosurface was protected by the construction of a closed building including visitor platforms and trails. The visible sauropod tracks are coloured for better visibility. This site is a designated German National Monument and it is part of the private "Dinosaurierpark Münchehagen" that is built around the tracksite (Fischer and Thies 2000). (**g**) Dinotec tracksite in Porrentruy (Canton Jura, NW Switzerland) provides another example of a different way to preserve tracks (Comment and Paratte 2013; Marty et al. 2013)

Fig. 4.7 Overview of The Devil's Tracks at Roccamonfina (Italy; Avanzini et al. 2008). (**a**) The tracks are preserved in an indurated pyroclastic flow. The site topography is the primary slope down which the track-maker descended. As shown the site has a gated metal walkway and display boards in public areas. It is in the care of the local community. (**b**) Close up of one of the trackways showing the zigzagged decent of the track-maker

apply to human tracksites where the substrate is usually much more erodible due to their relatively young age. Exposed display (Fig. 4.5a) is not an option at most human tracksites although there are a few exceptions, like the tracks at Roccamonfina in Italy which are preserved in an indurated pyroclastic flow (Avanzini et al. 2008) and are open to the air with a metal walk-way along the top of the site to provide a viewing platform (Fig. 4.7). In practice at most soft-sediment sites this is just not practical due to the erosive power of rainfall and/or run-off, especially in seasonal climates where it may be coupled with seasonal desiccation and the spalling of individual grains. Surface hardening of exposed sediment with resin (e.g., Agnew et al. 1989) might hinder erosion but it is hard to envisage stabilisation of the whole surface without causing changes to the moisture content and dynamics of underlying or adjacent beds leading to structural failure of the hardened slab.

Construction of a shelter to protect the tracked surface from the elements (Fig. 4.6; Demas 2002), such as that adopted at the Acahualinca footprint site in Nicaragua (Schmincke et al. 2010) along with control of surface run-off may provide an option. Agnew (2001) provides a useful review of the factors to be considered when opting for some form of shelter at archaeological sites and suggests that solutions require high capital investment, on-going maintenance, a commitment from the local community and the availability of tourists. It is an approach that has been followed for a number of dinosaur tracksites (Fig. 4.5). The 'value' of a site, whatever the combination of metrics used to evaluate this has to be deemed sufficient to merit such investment and the engagement of local stakeholders is critical to the success of such ventures. The Acahualinca site in Nicaragua works because of the proximity to the capital Managua. The final on-site option is buried display in which the site is covered over and protected for the long-term using the lessons gained from Laetoli (Musiba et al. 2008) which when coupled with either on-site display boards or an exhibit in a local museum does, if executed well, provide a viable conservation strategy. The loss of the site to the scientific and local community may however be significant here and constant re-excavation is not really an option as it would most certainly create some form of physical damage. There has to be a game changer in terms of new research techniques to justify this and then the question that remains is who gets to decide this and by what criteria?

Off-site conservation options involve such things as the removal of individual blocks or sections of a surface to either a local or regional/national museum for display. Both the South African tracksites of Langebaan and Nahoon have been removed as blocks and are now stored in museums (Roberts 2008; Fig. 4.8), although their removal has been facilitated by the material being heavily lithified and therefore able to withstand block removal. The importance of maintaining site provenance is really important however. Two human tracks from Cuatro Ciénegas (Mexico) are stored in the Museum of the Desert at Saltillo (Fig. 4.9), but the connection with the original site was lost until it was recently rediscovered or given the recent dating perhaps not (Gonzalez et al. 2007, 2009; Felstead et al. 2014). The site provides a cautionary tale about breaking the linkage between the provenance of a track and the original site and the importance of thorough documentation when excavating. While it is no doubt possible to first harden and then remove blocks of soft-sediment the ethics of doing so, even when a site may be lost to erosion, is questionable and the long-term preservation of individual tracks is not necessarily desirable because it is an assemblages of tracks and trackways as a whole that is usually of scientific importance. Large areas of soft-sediment however even if hardened are unlikely to survive block removal and transport. Linked to this are more traditional options of casting using latex or other casting media (Fig. 4.10). The success of the Laetoli casts (Figs. 3.4a, b and 3.5) illustrates that this can work and extensive experience exists around this (Leakey and Harris 1987). The problem here

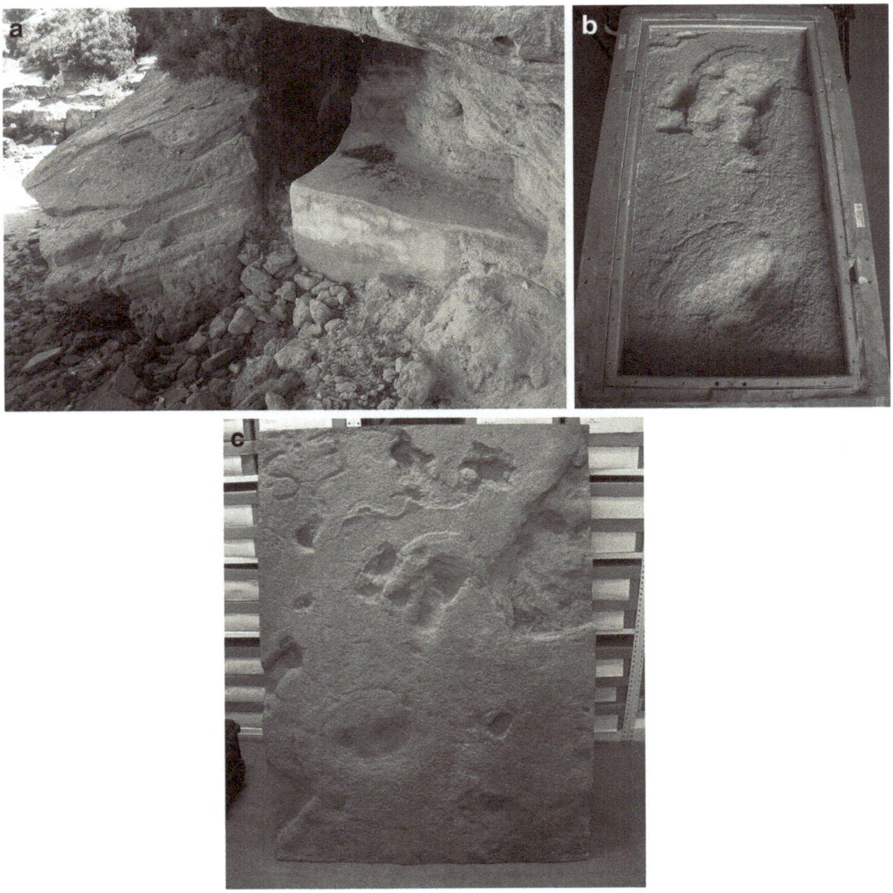

Fig. 4.8 Photographs of the human tracks from Langebaan in South Africa, both of which have been removed for different reasons to local museums (Roberts 2008). (**a**) The actual site at Langebaan from which the tracks have been removed. Note the carved graffeti on the site, which was one of the primary conservation threats. (**b**) The actual block removed, now encased in metal trolley for ease of movement at the Iziko South African Museum in Cape Town. (**c**) Cast of the tracks taken prior to the block removal. Cast held at Iziko South African Museum in Cape Town

is that the potential to damage soft-sediment sites in doing so is a considerable challenge. There is also the question of what should and should not be cast and the associated sampling strategy may have important implications for future use of a site within the research community. For example, the prints along the Sefton Coast have been preserved by a number of individuals over the years using Plaster of Paris; the fact that this may be destructive is beside the point since the prints are quickly lost to coastal erosion. A large collection of casts exist in various private and public collections, but whole trackways don't as the sampling tends to focus on

Fig. 4.9 Picture of the Cuatro Ciéngas tracks held at the Museum of the Desert at Saltiho in northern Mexico. These tracks have been recently dated by Felstead et al. (2014) and are older than the tracks recently re-discovered at Cuatro Ciéngas (Gonzalez et al. 2009). See Fig. 3.11 for images of the actual field site

one-off examples often with little associated documentation. They do provide as all casts do, however, a resource and opportunity for public display and engagement.

The final option is to do nothing, but to put in place an on-going monitoring and recording programme consistent with, for example, approaches within rescue archaeology. This is strongly advocated by Bennett et al. (2013) as the only viable option for the majority of soft-sediment sites not only for those like the tracks of the Sefton Coast where continual exposure is the order of the day but also for sites like Ileret. While each track at a site like Ileret may be highly valuable the very act of continued slope erosion will reveal more of the imprinted surface and their continual erosion is therefore not without positive benefits provided that the data is captured and recorded systematically as it is exposed. Bennett et al. (2013) acknowledged the futility of trying to preserve soft-sediment sites such as that at Ileret and argued that the emphasis

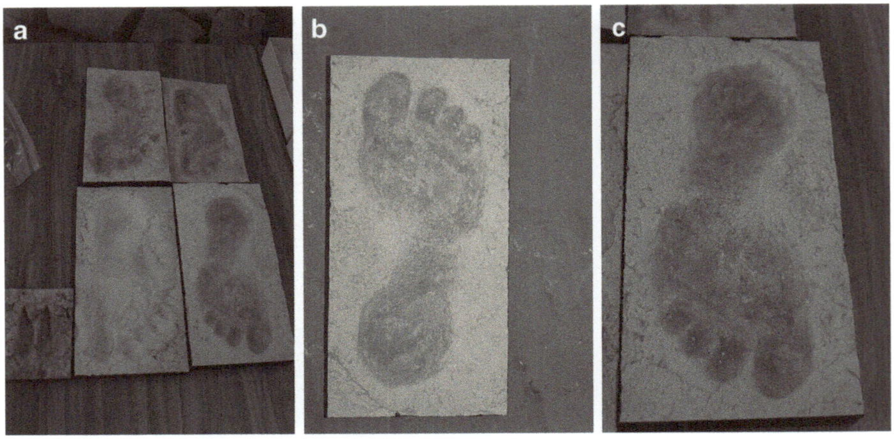

Fig. 4.10 A selection of plaster casts made from latex moulds held at the Museum of the Desert at Saltiho in northern Mexico from various sites including Cuatro Ceingas and Acahualinca. Note both the bulk and the lack of context. The area of the true track has been coloured slightly to enhance their appearance

should change from one of conservation of the physical tracks to the more abstract conservation of scientific data and in ensuring its subsequent open and public dissemination for the benefit of all stakeholders, not just those who collected it. The key to such a strategy is how the data is recorded, the associated quality, and how it is then shared throughout the scientific community for use by all, as well as how this resource is used to drive in-country tourism activity.

Bennett et al. (2013) argued that the conservation strategy for human tracksites falls into two broad categories, which grade one to the other, namely 'record and rescue' and 'site preservation' (Fig. 4.3). They argued that the only option in recording a site is for the digital capture of prints as set out in Sect. 2.4 and that this binary approach to site conservation occurs irrespective of 'value' whether that is scientific or on some other basis. It is worth drawing attention here to the similarity with dinosaur tracksites found in the floors of working quarries which are due for infill with landfill or continued quarrying, here 'record and rescue' approaches have been adopted to good effect (e.g., Day et al. 2004). This framework provides an interesting perspective on the Ileret tracks since one might at first assume that conservation of the artefact should be the key priority given their potential scientific importance. However, at sites like this where there are extensive areas with the *potential* to contain prints, in this case widespread areas of a fine-grained over-bank deposits with evidence of hominin congregation, then one could argue that there is a high probability that tracks are more ubiquitous than previously thought, but at this point are just undiscovered. This is confirmed by the fact that further exposures of animal prints have been found laterally to the Koobi Fora GaJi10 site (see Sect. 3.1.3) and additional hominin prints at Ileret FwJj14E. The limitation to their study, especially at Ileret, is actually that they are buried deeply by overburden, making continued erosion a positive factor since it will reveal new prints for study. This is similar to the UK coastal tracksites in the Severn Estuary (Aldhouse-Green et al. 1992) or on the Sefton Coast (Roberts et al. 1996) where continual coastal erosion exposes new prints for study which cannot be preserved physically only recorded and rescued. The key difference here is that unlike a typical UK coastal fossil site the value to the palaeoanthropological community of a site such as Ileret is much greater as is the symbolic and cultural value to both indigenous and foreign populations. In both cases we would argue that it is the quality of the record and digital rescue approach used that is key. It is worth noting that the stakes are potentially high if one gets this 'record and rescue' approach wrong as arguably occurred at Happisburgh (Ashton et al. 2014) given the failure to acquire high quality digital elevation models before the prints were lost to erosion, although we acknowledge the difficulty of doing this if the tracks are constantly infilling with water.

4.5 Summary

Considering both the short- and long-term conservation options is something that every geoarchaeologist who works at a new human tracksite must consider. The excavation strategy, methods of documenting and digital data recording of the tracks all have implications for how a site can or cannot be conserved in some way in the long-term. The temptation is for the field scientist to simply 'grab' their data and exit rather than to consider their role in the conservation of a site. There are notable exceptions this and this chapter hopefully provides some information to help geoarchaeologists to at least consider how to discharge the duty of stewardship that goes with discovery. The answers are often not easy and the options limited especially at vulnerable soft-sediment sites. In most cases we would argue that this conservation strategy is likely to focus around 'record and digitally rescue' and the aim should be about ensuring that the resources are freely available to create at some point in the future a virtual representation of the site, along with careful documentation, and if appropriate sample archiving, of the sedimentary and palaeoenvironmental context so that a potentially timeless resources is available for future scientific study. This resource needs to be openly accessible to all and crucially provide a platform for public engagement both at and beyond the site. Even if the tracks are naturally eroded or removed by scientists, the context is very valuable both to scientists and the community; therefore appropriate signposting by local authorities is extremely important. It is also about timely and continual intervention to make sure data is not lost, something that needs to involve the whole of the scientific as well as the local community in that process. The revolution under way in digital photogrammetry (see Sect. 2.4) has the potential to involve a wider cross-section of people in the capture of that data as it first becomes exposed. For example, you can envisage a radical strategy at a site such as that on the Sefton Coast in which it is the local amateurs and dog walkers who capture the images needed for photogrammetry with a bit of awareness and training. Such a strategy would truly be a pioneering piece of participative science and might provide a very rewarding data base of information over time, far more than that which could be collected by the occasional visiting scientist.

References

Agnew N (2001) Methodology, conservation criteria and performance evaluation for archaeological site shelters. Conserv Manag Archaeol Sites 5:7–18

Agnew N, Demas M (1998) Preserving the Laetoli footprints. Sci Am 279:44–55

Agnew N, Demas M (2004) Monitoring through replication: design and evaluation of the monitoring reburial at the Laetoli trackway site. Conserv Manag Archaeol Sites 6:295–304

Agnew NH, Griffin H, Wade M, Tebble T, Oxman W (1989) Strategies and techniques for the preservation of fossil tracksites. In: Gillette DD, Lockley MG (eds) Dinosaur tracks and traces. Cambridge University Press, Cambridge

Aldhouse-Green SHR, Whittle AWR, Allen JRL et al (1992) Prehistoric human footprints from the Severn Estuary at Uskmouth and Magor Pill, Gwent, Wales. Archaeol Cambrensis 149:14–55

Ashton N, Lewis SG, De Groote I et al (2014) Hominin footprints from early Pleistocene deposits at Happisburgh, UK. PLoS One 9(2):e88329

Avanzini M, Mietto P, Panarello A et al (2008) The Devil's trails: middle Pleistocene human footprints preserved in a volcanoclastic deposit of Southern Italy. Ichnos 15:179–189

Beckerman W, Pasek J (2001) Justice, posterity and the environment. Oxford University Press, Oxford

Behrensmeyer AK, Laporte LF (1981) Footprints of a Pleistocene hominid in northern Kenya. Nature 289:167–169

Bennett MR, Doyle P (1997) Environmental geology. Wiley, Chichester

Bennett MR, Harris JWK, Richmond BG et al (2009) Early hominin foot morphology based on 1.5 Million year old footprints from Ileret, Kenya. Science 323:1197–1201

Bennett MR, Falkingham P, Morse SA et al (2013) Preserving the impossible: conservation of soft-sediment hominin footprint sites and strategies for three-dimensional digital data capture. PLoS One 8(4):e60755. doi:10.1371/journal.pone.0060755

Bruschi VM, Cendrero A, Albertos JAC (2011) A statistical approach to the validation and optimisation of geoheritage assessment procedures. Geoheritage 3(3):131–149

Castanera D, Barco JL, Diaz-Martinez I et al (2011) New evidence of a herd of titanosauriform sauropods from the Lower Berriasian of the Iberian range (Spain). Palaeogeog Palaeoclimat Palaeoecol 310:227–237

Comment G, Paratte G (2013) Géologie et paléontologie en Ajoie: prospection dans des couches jurassiques. Actes 2013 de la société jurasienne d'émulation Actes 2013:17–46

Day MH, Wickens EH (1980) Laetoli pliocene hominid footprints and bipedalism. Nature 286:385–387

Day JJ, Norman DB, Gale AS et al (2004) A Middle Jurassic dinosaur trackway site from Oxfordshire, UK. Palaeontology 47:319–348

Demas M (2002) Annotated bibliography on protective shelters for archaeological sites. Conserv Manag Archaeol Sites 5:91–105

Demas M, Agnew N (2006) Decision making for conservation of archaeological sites: the example of the Laetoli hominid trackway, Tanzania. In: Of the past, for the future, integrating archaeology and conservation Getty conservation institute symposium proceeding publication of the 5th World Archaeological Congress, Getty Publications, Washington, DC, 22–26 June 2003

Diedrich C (2010) Upper Jurassic tidal flat megatracksites of Germany – coastal dinosaur migration highways between European islands, and a review of the dinosaur footprints. Palaeobiodivers Palaeoenviron 91:129–155

Dingwall HL, Hatala KG, Wunderlich RE et al (2013) Hominin stature, body mass, and walking speed estimates based on 1.5 million-year-old fossil footprints at Ileret, Kenya. J Hum Evol. doi:10.1016/j.jhevol.2013.02.004
Dowling RK (2011) Geotourism's global growth. Geoheritage 3:1–13
Doyle P, Bennett MR (1998) Earth heritage conservation: past, present and future agenda. In: Bennett MR, Doyle P (eds) Issues in environmental geology: a British perspective. Geological Society of London, London, pp 41–67
Eder W, Patzak M (2004) Geoparks – geological attractions: a tool for public education, recreation and sustainable development. Episodes 27(3):162–164
Falk A, Hasiotis ST, Martin LD (2010) Feeding traces associated with bird tracks from the Lower Cretaceous Haman Formation, Republic of Korea. Palaios 25:730–741
Farsani NT, Coelho COA, Costa CMM, Amrikazemi A (2014) Geo-knowledge management and geoconservation via geoparks and geotourism. Geoheritage. doi:10.1007/s12371-014-0099-7
Fassoulas C, Mouriki D, Dimitriou-Nikolakis P et al (2012) Quantitative assessment of geotopes as an effective tool for geoheritage management. Geoheritage 4(3):177–193
Felstead NJ, Gonzalez S, Huddart D (2014) Holocene-aged human footprints from the Cuatrociénegas Basin, NE Mexico. J Archaeol Sci 42:250–259
Fischer R, Thies D (2000) Das Dinosaurier-Freilichtmuseum Münchehagen und das Naturdenkmal "Saurierfährten Münchehagen", 2nd edn. Rehburg-Loccum. Münchehagen, 64 pp
Gonzalez AH, Lockley MG, Rojas CS et al (2007) Notes on the re-discovery of a 'lost' hominid footprint site from the Cuatro Cienegas basin (Coahuila), Mexico. N M Mus Nat Hist Sci Bull 42:11–15
Gonzalez AH, Lockley MG, Rojas CS et al (2009) Human tracks from Quaternary tufa deposits, Cuatro Ciénegas, Coahuila, Mexico. Ichnos 16:12–24
Gray M (2002) Geodiversity. Wiley, Chichester
Henriques MH, dos Reis RP, Brilha J, Mota T (2011) Geoconservation as an emerging geoscience. Geoheritage 3(2):117–128
Hose TA (2012) 3G's for modern geotourism. Geoheritage 4:7–24
Huh M, Paik IS, Lockley MG et al (2006) Well-preserved theropod tracks from the Upper Cretaceous of Hwasun County, southwestern South Korea, and their paleobiological implications. Cretac Res 27:123–138
Hwang K-G, Huh M, Lockley MG et al (2002) New pterosaur tracks (Pteraichnidae) from the Late Cretaceous Uhangri Formation, SW Korea. Geol Mag 139:421–435
Hwang K-G, Lockley MG, Huh M et al (2008) A reinterpretation of dinosaur footprints with internal ridges from the Upper Cretaceous Uhangri Formation, Korea. Palaeogeog Palaeoclimat Palaeoecol 258:59–70
Kaever M, Lapparent AF (1974) Les traces de pas de dinosaures du Jurassique de Barkhausen (Basse Saxe, Allemagne). Bull Soc Géol France 16(7):516–525
Kiernan K (2013) The nature conservation, geotourism and poverty reduction nexus in developing countries: a case study from the Lao PDR. Geoheritage 5(3):207–225
Kim BS, Huh H (2010) Analysis of the acceleration phase of a theropod dinosaur based on a Cretaceous trackway from Korea. Palaeogeog Palaeoclimat Palaeoecol 293:1–8
Kim JY, Lockley MG (2012) New Sauropod tracks (*Brontopodus pentadactylus ichnosp.* nov.) from the Early Cretaceous Haman Formation of Jinju area, Korea: implications for Sauropods Manus morphology. Ichnos 19:84–92
Kim JY, Lockley MG, Seo SJ et al (2012) A paradise of Mesozoic birds: the world's richest and most diverse cretaceous bird track assemblage from the Early Cretaceous Haman Formation of the Gajin Tracksite, Jinju, Korea. Ichnos 19:28–42
Leakey MD, Harris JM (1987) Laetoli a Pliocene site in northern Tanzania. Clarendon/Oxford University Press, Oxford
Lee Y-N, Huh M (2002) Manus-only sauropod tracks in the Uhangri Formation (Upper Cretaceous), Korea and their paleobiological implications. J Paleont 76:558–564
Lockley MG, Meyer CA (1997) Sacred ground: walking in the footsteps of our ancestors. In: Huh M, Yang SSY, Lockley MG, Lee YN (eds) Proceedings of the international Dinosaur

symposium for the Uhangri Dinosaur Centre and Theme Park. Chonnam National University, Gwangju, pp 211–220
Lockley MG, Wright JL, Thies D (2004) Some observations on the Dinosaur tracks at Münchehagen (Lower Cretaceous), Germany. Ichnos 11:261–274
Lockley M, Roberts G, Kim JY (2008) In the footprints of our ancestors: an overview of the hominid track record. Ichnos 15:106–125
Lockley MG, Huh M, Kim BS (2012) Ornithopodichnus and pes-only sauropod trackways from the Hwasun tracksite, Cretaceous Korea. Ichnos 19:93–100
Macfarlane R (2012) The old ways: a journey on foot. Hamish Hamilton, London
Mampel L, Cobos A, Alcalá L, Luque L, Royo-Torres R (2009) An integrated system of heritage management applied to dinosaur sites in Teruel (Aragón, Spain). Geoheritage 1(2–4):53–73
Marty D, Cavin L, Hug WA et al (2004) The protection, conservation and sustainable use of the Courtedoux dinosaur tracksite, Canton Jura, Switzerland. Revue de Paléobiologie volume spéciale 9:39–49
Marty D, Waite R, Hug WA et al (2013) Fieldtrip to the Late Jurassic (Kimmeridgian) CPP dinosaur tracksite and nearby reference section Roche de Mars (Porrentruy, Ajoie district, Canton Jura). In: Koppka J, Hostettler B, Marty D et al (eds) Exkursionsführer der Jahrestagung 2013 in Porrentruy (Kanton Jura, Schweiz) der Subkommission für Jurastratigraphie der Deutschen Stratigraphischen Kommission (DSK), pp 3–22
Matthews TJ (2013) Integrating geoconservation and biodiversity conservation: theoretical foundations and conservation recommendations in a European Union context. Geoheritage 6(1):1–14
Mayor A, Sarjeant WAS (2001) The folklore of footprints in stone: from classical antiquity to the present. Ichnos 8:143–163
Morse SA, Bennett MR, Liutkus-Pierce C et al (2013) Holocene footprints in Namibia: the influence of substrate on footprint variability. Am J Phys Anthropol 151:265–279
Musiba C, Mabulla A, Selvaggio M, Magori C (2008) Pliocene animal trackways at Laetoli: research and conservation potential. Ichnos 15:166–178
Page KN (1998) England's Earth heritage resource – an asset for everyone. In: Hooke J (ed) Coastal defence and earth science conservation. The Geological Society, London, pp 196–209
Pérez-Lorente F, Cuenca-Bescós G, Aurell M et al (1997) Las Cerradicas tracksite (Berriasian, Galve, Spain): growing evidence for quadrupedal ornithopods. Ichnos 5:109–120
Reis RP, Henriques MH (2009) Approaching an integrated qualification and evaluation system for geological heritage. Geoheritage 1(1):1–10
Roberts DL (2008) Last interglacial hominid and associated vertebrate fossil trackways in coastal eolianites, South Africa. Ichnos 15:190–207
Roberts G, Gonzalez S, Huddart D (1996) Inter-tidal Holocene footprints and their archaeological significance. Antiquity 70:647–651
Santos VF, Marques de Silva C, Rodrigues LA (2008) Dinosaur track sites from Portugal: scientific and cultural significance. Oryctos 8:77–88
Schmincke HU, Rausch J, Kutterolf S et al (2010) Walking through volcanic mud: the 2,100 year-old Acahualinca footprints (Nicaragua) II: the Acahualinca people, environmental conditions and motivation. Int J Earth Sci 99:279–292
Strüm B (1994) The geotope concept. In: O'Halloran D, Green C, Harley M, Stanley M, Knill J (eds) Geological and landscape conservation. The Geological Society, London, pp 27–31
UNESCO (2006) UNESCO Geoparks Programme Feasibility study. UNESCO, Paris
Wilson C (1994) Earth heritage conservation. Geological Society of London/Open University, Milton Keynes
Wimbledon WAP (1996) GEOSITES – a new conservation initiative. Episodes 19:87–88
Wimbledon WAP (2006) Commercial exploitation versus sustainable use. Earth Herit 27:7–8

Chapter 5
The Role of Substrate in Track Formation and Topology

Abstract Before any inferences about a human track-maker can be made from the tracks they leave we first need to understand how substrate properties (consistency) mediate and record the interaction between the foot and the ground. A substrate has a profound influence in a range of different ways: it controls the way in which a track-maker walks; it moderates and cushions the pressure interaction between the sole and the substrate and therefore the depth record that results; it determines how the strain is accommodated and the type of deformation that occurs; and it determines the immediate survival and preservation of a track as well as its longer term taphonomy. We argue that geology matters in the study of human tracks and in this chapter we explore this first by looking at the role of substrate as a variable in track formation and secondly by looking at the taphonomic process which bear on track formation and preservation in the geological record.

5.1 Substrate Controls: Introduction

A sediment's consistency has the potential to influence track topology, the level of anatomical detail cast and the susceptibility of a track to taphonomic modification (e.g., Scrivner and Bottjer 1986; Cohen et al. 1993; Allen 1997; Gatesy et al. 1999; Diedrich 2002; Fornós et al. 2002; Melchor et al. 2002, 2006; Manning 2004; Milàn 2006; Scott et al. 2008, 2010; Jackson et al. 2009). If one walks on coarse gravel the sediment may be indented with the outline of the foot but little anatomical detail will be preserved, while in contrast a soft clay may preserve not only the callouses but the friction ridges and creases of the foot (Lockley 1991; Morse et al. 2013). Small-scale variations in sediment consistency may also lead to variation in track topology along a trackway and must of course be accounted for in making comparisons between tracks formed in different depositional settings (Morse et al. 2013). More importantly a substrate may influence gait, for example one may walk differently one a slippery surface than on a rough one where footing is secure (Cham and Redfern 2002) and the energy cost of walking on different substrates also varies (Crevier-Denoix et al. 2010). Sediment consistency and the pertaining environmental conditions will also control the taphonomic changes that a track undergoes during burial and/or digenesis (Laporte and Behrensmeyer 1980; Marty et al. 2009; Scott et al. 2009, 2010).

M.R. Bennett, S.A. Morse, *Human Footprints: Fossilised Locomotion?*,
DOI 10.1007/978-3-319-08572-2_5

The importance of understanding the role of substrate and of environmental conditions more generally is given added impetus by the fact that the number of hominin tracksites around the world is limited and therefore comparisons between sites with different sedimentological and palaeoenvironmental contexts is a necessity, although not ideal, if inter-species comparisons are to be made (e.g., Charteris et al. 1981, 1982; Tuttle 1990; White and Suwa 1987; Meldrum 2004; Bennett et al. 2009; Meldrum et al. 2011; Crompton et al. 2012). Central to any biomechanical inferences made is our understanding of the way in which track depth, and its relative distribution within a track, is correlated in some way with the dynamic foot pressures of the track-maker and by inference the motion of their limbs (D'Aout et al. 2010; Bates et al. 2013). Vertebrate ichnology has a head start in exploring some of these issues, due to the great number of known dinosaur tracksites (e.g., Allen 1997; Gatesy et al. 1999; Diedrich 2002; Fornós et al. 2002; Melchor et al. 2002, 2006; Manning 2004; Milàn 2006, Milàn and Bromley 2008; Falkingham et al. 2010, 2011) and a number of modern analogue (neoichnology) and laboratory studies have been conducted to explore the role of substrate and taphonomy (e.g., Buckland 1928; Padian and Olsen 1984; Demathieu 1987; Farlow 1989; Brand 1996; Davis and Stevenson 2007; Genise et al. 2009). In contrast the particular studies pertaining to human tracks are more limited and understanding of the role of substrate is only just emerging (Marty et al. 2009; Bates et al. 2013; Morse et al. 2013). In exploring this we will first introduce a simple model of human track formation, before looking in turn at the influence of sediment properties and taphonomy.

5.2 Models of Human Track Formation

As a foot makes contact with a sediment surface there is a transfer of force (pressure) to the substrate which is then deformed if the applied force exceeds some measure of the substrate's strength and that there is no elastic response. There are three principal components to the creation of a track, essentially to the creation and retention of a surface void (Fig. 1.2): (1) the irreversible compression of sediment below the plantar surface of the foot, effectively providing thorough consolidation space for the foot; (2) sediment displacement (i.e., transport through failure and deformation) below and away from the foot from areas of high pressure, to adjacent areas of lower pressure; and (3) the physical excavation and removal of sediment through plantar shear beneath the sole of the foot or by adhesion to the foot. None of these mechanisms are mutually exclusive and some combination will hold in the formation of most tracks.

Figure 5.1a shows a track formed by simple compression; note the extensional tension fractures around the sides and the compressed plantar sediment layers. The degree to which the applied stress can be accommodated solely by compression will depend on such sedimentary properties as grain-size, sorting, grain shape, porosity, packing, consolidation and pore-water content all of which will control the degree

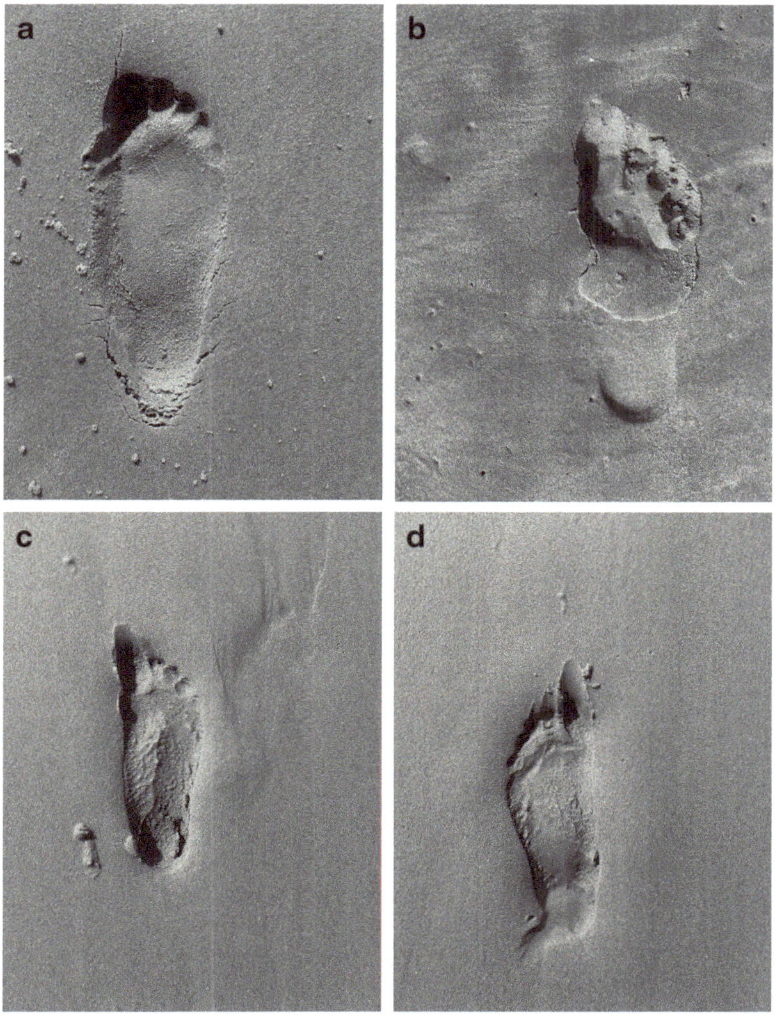

Fig. 5.1 A series of modern human tracks made on a beach at Bournemouth in the UK. (**a**) Track formed by simple compression note the concentric and radial extensional fractures around the margins. (**b**) Note the movement of a thin slab of sediment within the floor of the track, caused by rotation of the ball of the foot. (**c**) Note the longitudinal ridge in the base of the track caused by the lifting of damp sediment adhering to the sole of the foot. (**d**) Note the elongated first toe caused by forward drag and the failure of the heel wall

of strain compression that can be accommodated (Allen 1997; Craig 2004). These properties are broadly captured by the sediment's natural bulk density which is effectively the mass of an intact volume (Manning 2004). Sediment that is poorly consolidated has a low bulk density and will be able to accommodate much more compressive strain than a sediment with a higher bulk density and consequently less void or pore-space to accommodate compression as the individual grains move closer. In general terms a sediment's compressive strength will increase with its bulk density (Craig 2004) and with its moisture content up to a critical hydraulic threshold (Manning 2004). Dry sand has little strength compared to one which is damp, since in damp sediment, water coats the grains and the surface tension of the water helps bind the grains together. As the water content increases however and the air is driven from the sediment's pores this water begins to push the grains apart and lubricate their relative motion, which reduces the sediment's strength. The water is effectively incompressible when loaded and unless it can drain to areas of lower pressure and it will cause the sediment to ultimately fail. The rate at which the stress is applied, the porosity and permeability of the sediment as well as its vertical variation are all critical at this point in determining whether and how the sediment will fail (Craig 2004). For example, rapid footfall does not always provide sufficient time for pore-waters in sediment with a low permeability to drain and the sediment may fail rather than compress, where a similar pattern of footfall may cause compression if the sediment had a higher permeability allowing pore-waters to drain more effectively.

The void necessary for a track may also be formed through sediment displacement (failure) either because simple compression is unable to accommodate the strain adjustment necessary or because the rate of stress application is too rapid for that adjustment to occur. At its simplest displacement occurs around the entire margin of a track forming an elevated rim structure (Figs. 5.2 and 5.3; Thulborn 1990) however in other cases it may become associated with a particular horizontal force vector which concentrates deformation in a particular quadrant of a track (Brown 1999). In fine-grained sediment with a high mud fraction (<µ63 m) deformation will tend to occur plastically assuming that the pore-water contents lies somewhere between the plastic and liquid limits (Craig 2004). In damp sand deformation may occur in a brittle rather than ductile fashion and involve the movement of thin sediment slabs, a mechanism which grades into the final track formation process (Fig. 5.4; Appendix, Track 4).

The third mechanism involves the physical excavation of material lifted from the track by the foot either to the front or rear. Such ejecta are associated with either sediment adhering to the sole of the foot or being flicked out by the action of the toes (Bird 1944; Allen 1997). Consolidation of sediment beneath the plantar surface of a track may also lead to the formation of a shear zone between the sole of the foot and the consolidating tracking surface beneath, such that thin slivers of sediment move in an anterior direction along the plantar surface of the track (Figs. 5.1b, 6.11 and 6.12; Appendix, Tracks 8, 10 and 11). In the extreme this process grades naturally into the process of block displacement described above as the movement exceeds the confines of the plantar surface (Fig. 5.4). This is particularly common in less ductile material especially slightly damp sand with good inter-granular capillary action. Note that the rate of applied stress may be significant here; high strain rates are often associated with more brittle deformation, which is one of the reasons why brittle deformation is a more prominent feature of tracks left by faster rather than slower walking (see Sect. 6.4). It is important to recognize that the movement of

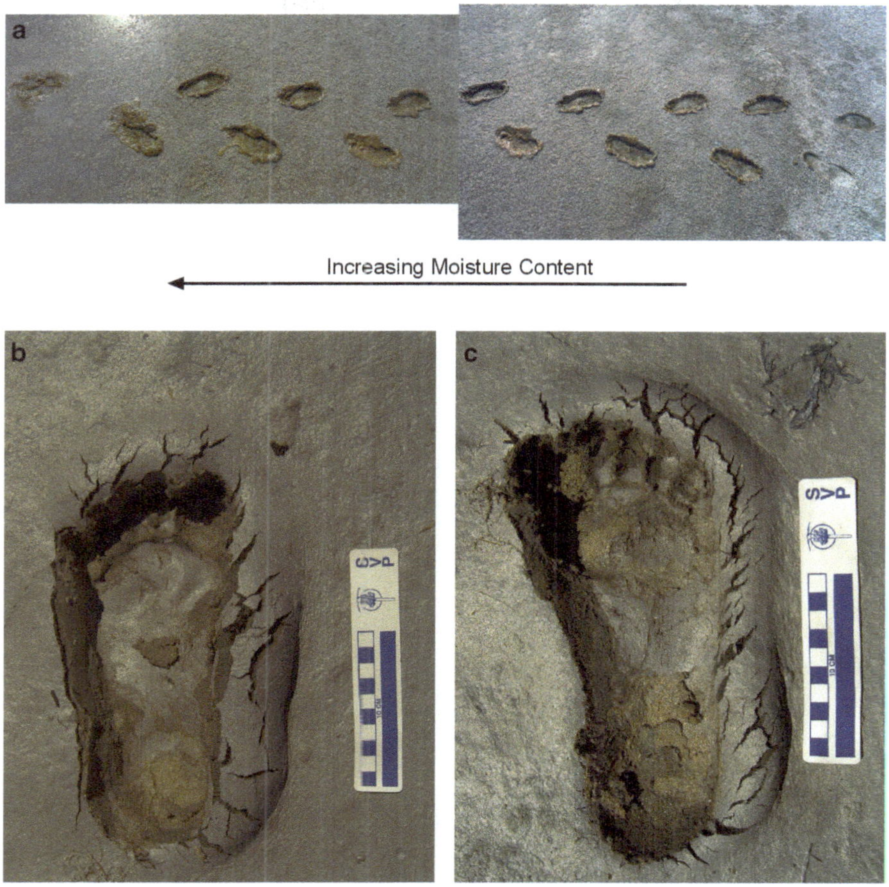

Fig. 5.2 A series of images showing various aspects of substrate. (**a**) Trackway emerging from shallow water showing the transition in typology with decreasing water content. (**b, c**) Two images showing rim structures associated with sediment displacement, note the cracking of the rims following deformation (All these images are kindly provided by Daniel Marty and reproduced here with kind permission)

sediment within a track may obscure much of its anatomy and that the movement may not always be evident in all cases.

The presence or absence of rim structures (Bourrelet; Figs. 5.2 and 5.3) around the margin of a track provides some indication as to the relative importance of the above processes. Where rim structures are absent the void occupied by the foot must have been accommodated primarily by compression alone, although this does not mean that sediment transfer has not occurred as well, just that compression below the track and during lateral transfer is sufficient to accommodate the space occupied by the foot. It is also important to note that final track topology will be influenced by effects caused by foot withdrawal and associated drag as well as the collapse of

Fig. 5.3 Human tracks with rim structures made in laminated silt overlying a firm sub-base at Walvis Bay in Namibia see Sect. 3.2.3 for details

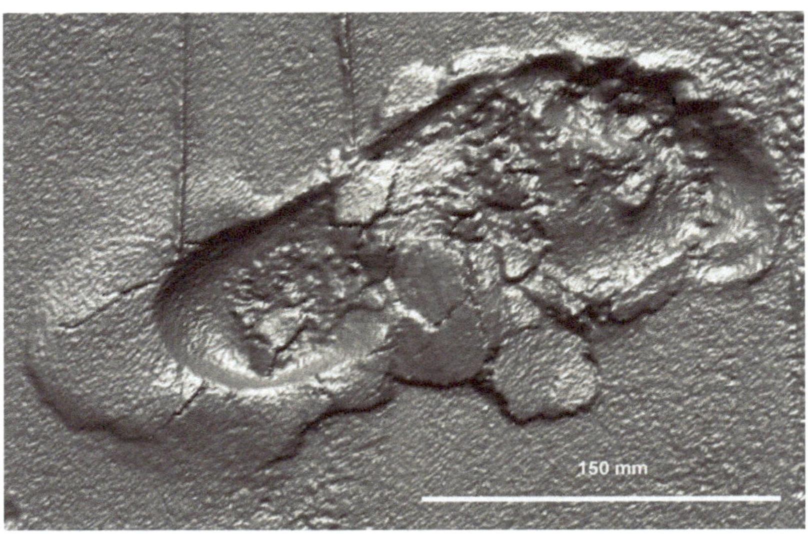

Fig. 5.4 Optical laser scan of a left human track made in soft sand. Note the thin slabs of sediment moved both within the sole of the foot in a posterior direction and medially. See also the broad rim structure around the heel and the radial fractures

the track walls as the substrate relaxes and rebounds. Foot withdrawal in deeper tracks is associated with the creation of suction below the sole of the foot which may cause a track's side walls to be drawn inward something which has been used to explain the very narrow tracks at Ileret for example (Bennett et al. 2009; Morse et al. 2013). The creation of rim structures may also lead to subsequent track-wall loading and closure as the walls buckle inwards and in the worst cases failure. On a small scale Fig. 5.1c illustrates the effects of plantar suction, while Fig. 5.1d show track modification due to both toe drag and a failure of the heel wall. The deeper the track, the taller the track walls, and therefore the more likely some form of track wall failure is; a problem which is often exacerbated by the collection of either surface or ground waters within a track.

Natural sediment is an admixture of particles of various sizes, air and pore-water and as such its description is complex (Craig 2004) and finding a suitable theoretical description is consequently difficult (Allen 1997). In the context of vertebrate ungulate tracks (Allen 1989, 1997) and those made by dinosaurs (Manning 2004) indenter theory has been applied and an elastic–plastic approximation of sediment rheology used to good effect. These studies emphasise sediment failure and use the theoretical pressure bulb and associated slip-line fields to understand the deformation below variously shaped indenters. Critical in these observations is the depth of the plastic layer (i.e. the presence or absence of a firmer sub-layer below) and the roughness of the indenter which all determine the lateral extent of deformation around the indenter and geometry of an un-deformed or 'dead' area beneath the indenter (Fig. 5.5a). The validity of this model as an approximation of sediment behaviour was confirmed by Allen's (1997) laboratory experiments using Plasticine rather than sediment and by direct field observations of ancient tracks exposed around the Severn Estuary. This theoretical and experimental approach was developed further by Manning (2004) in the context of dinosaur tracks and he recognised four failure modes (Fig. 5.5b) which equate in part to the compression and displacement model used above. In his 'general shear' scenario continuous failure surfaces develop around the edge of the indenter and following the classically defined Rankine shear zones (Craig 2004; Fig. 5.5a). In the case of 'local shear' there is significant compression under the indenter such that the failure surfaces don't reach the ground surface (Fig. 5.5b) as is the case in his third scenario (punctured shear). Manning's (2004) final failure mode recognises that an indenter may simply sink due to thixotropy if loaded rapidly and that track walls may flow inwards. The implication here is that while the surface topology of a track may be lost the plantar surface may be preserved at depth (Gatesy et al. 1999). The shape of an indenter has also been found to be important here and specifically the ratio of area to perimeter; the greater the perimeter length with respect to the area the less an indenter penetrates an elastic–plastic sediment (Falkingham et al. 2010). In the context of a human track this might explain why the hindfoot may penetrate further than the forefoot for a given applied pressure; in the former the heel has a simple outline, while in the latter the toes provide a more complex outline (Bates et al. 2013). It is also worth noting that further modelling by Falkingham et al. (2011) demonstrated that track formation was only possible over a limited range of loads effectively

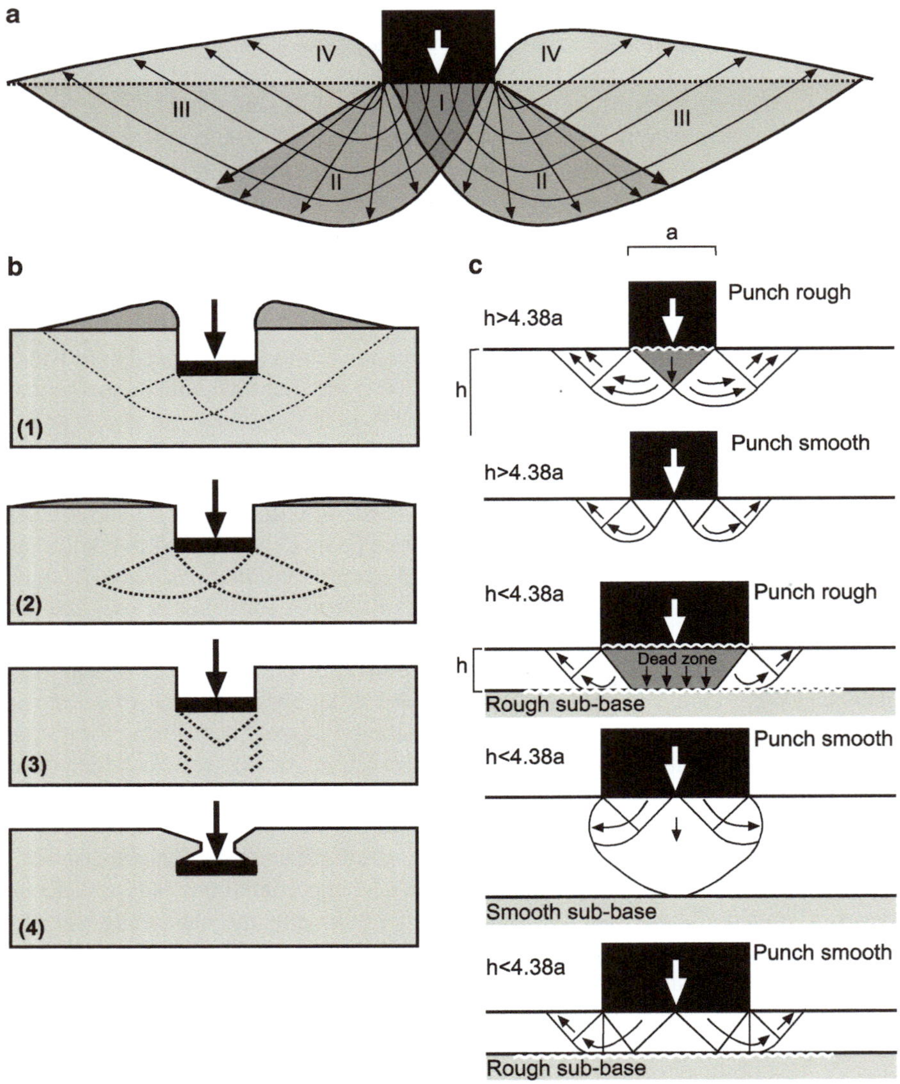

Fig. 5.5 Failure fields according to an elastic–plastic approximation associated with indenters of with different surface roughness. (**a**) General failure fields with associated with slip-line failure lines showing the relative position of the Rankine Shear Zones. The *arrows* depict the direction of sediment displacement: Zone I = Active Rankine Zone; Zone II = Radial Rankine Zone; Zone III = Passive Rankine Zone; and Zone IV = Displacement Rim (Modified from Manning (2004)). (**b**) Modes of failure according to Manning (2004): (1) = General Shear; (2) = Local Shear; (3) Punctured Shear; and (4) Liquefaction Failure. Failure/shear surfaces are shown by the *dotted lines*. (**c**) Failure modes associated with various indenters and sediment properties following Allen (1997). Note the 'dead zone' which occurs in certain cases below the indenter and the importance of depth to the rigid sub-base and the various patterns of failure/shear

introducing a preservation bias within some substrates which they referred to as the 'Goldilocks Effect'. Effectively the substrate acts as a sampling filter on what animals (applied loads) result in track preservation. They also noted that track depth was indicative of the thickness of the mechanically distinct sub-layers introduced into their model at various points, reinforcing the importance of vertical and spatial variation in substrate properties with respect to track formation.

5.3 Substrate Controls

There is a consensus within the literature on animal tracks that moisture content is critical across a range of grain size mixtures; too little and the tracks have poor definition, too much and they are lost as the sediment collapses. For example, Tucker and Burchette (1977) suggested that the presence of 20 % mud was important for track preservation. Scrivner and Bottjer (1986) documented the variation in Neogene avian and mammal track morphology in Death Valley (California) and found variations in moisture content gave rise to a range of track morphologies for a given track-maker. Moist or slightly damp sand was found to give the best preservation, while too much moisture lead to a situation where sediment either adhered to the foot creating a 'messy track' or collapse of the track-wall obscured the anatomical form of the track-maker. This is intuitive when one thinks about the tracks one leaves on a beach. If you walk close to the surf the tracks you leave are rarely held for long by the sand, collapsing almost as soon as they are formed by the flow of saturated sand. Just beyond the swash line however the sand is firm and your tracks are usually shallow – the firmest walking is usually had here. The sediment has high moisture content, is undisturbed and consequently the combination gives it a high bulk density, resistant to compression. As you move further up the beach, away from the swash line one's tracks get deeper as the surface layer of sand has begun to dry out and is more compressible. Your tracks are deeper with your foot pushing down to a firmer, damper sub-base of sand. The dry surface sand often moves in a zone of shear between the plantar surface of the foot and the damp sub-base. If one progresses still further up the beach the dry surface layer becomes much deeper and the sand is more disturbed, re-worked by the wind and by previous footfall, such that it does not hold tracks well and those that do form tend to be shallow, broad craters. We have just described the variation in bulk density that occurs with moisture content and sediment re-working, the compact zone near the swash line corresponds to a peak in bulk density associated with the critical hydraulic threshold described by Manning (2004). It is important to note that this pattern is different where high levels of mud (silt + clay) are present, here sediment failure dominates over compression and the decline in moisture away from the water line leads to firmer substrates as the decrease in moisture content enhances the level of cohesion present.

As one would expect a number of studies have stressed the proximity to a former or current shoreline in determining the moisture content and track topology (Van der Lingen and Andrews 1969; Lockley 1986; Zhang et al. 1998). The point is very well made in Fig. 5.2 which shows a trackway emerging from shallow water to dry land. Perhaps the classic study here is that of Cohen et al. (1993) who recognised a series of footprint zones around Lake Manyara in Tanzania. These consisted of a landward zone of firm, dry, salt encrusted sediment where only large ungulate tracks were found; lighter animals were unable to load the surface sufficiently to leave tracks. These tracks had a high survival potential compared to the inner strandline zone. Here the damper sediment was ideal to record a range of animal and bird tracks with excellent morphological detail, but the tracks had poor survival potential due to regular sediment re-working and animal trampling along the water edge. Their third zone (subaqueous zone) was characterised only by the tracks of larger mammal and the preservation potential was poor due to liquefaction of the saturated sediment. In a wonderful study of modern bird tracks Genise et al. (2009) recognised the importance of critical moisture content in optimising track morphology. Milàn (2006) came to similar conclusions in a study of modern emu tracks as analogues for dinosaur trackways. Scott et al. (2010) recognised not only the importance of moisture but also clay mineralogy and a site's overall geohydrology in determining the role of salt growth both in track formation and their subsequent taphonomy. These observations and the modelling work all suggests that vertical stratigraphy may also be of crucial importance in determining a track's topology (Allen 1997; Manning 2004; Falkingham et al. 2011).

With the exception of a few recent studies the influence of substrate on human tracks has not been explored in detail despite the wide range of environments from which they have been recovered (Lockley et al. 2008; Table 3.1). Before examining the 'notable exceptions' in detail it is perhaps worth first looking at tracks made at different locations each with varying combinations of grain-size and moisture content, in a single natural depositional laboratory in this case the Conwy Estuary in North Wales (Fig. 5.6a). At each site the senior author made a series of 15–20 tracks walking at a normal and even pace. Tracks were recorded using photogrammetry following Falkingham (2012), along with sediment grain-size and moisture content determinations for each site following standard laboratory procedures.

The most homogenous as well as mud- and water-rich sediment is found at Site A (Fig. 5.6; 63 %=silt+clay; 81 %=water) and resulted in deep tracks (100–200 mm) typically with a 'wedge-shaped' longitudinal cross-section, in which the ball and toes are much deeper than the heel (Figs. 5.6b and 5.7). The toes are characteristically splayed and in particular the first toe is widely abducted (Fig. 5.7a) reflecting the forced separation of the toes by sediment both during indenting and withdrawal. Broad rim structures are present due to both sediment displacement and sediment evacuation during foot withdrawal (Fig. 5.7b, c). Radial fractures open up transverse to the central axis of the rim and the track walls are seen to close over time due to loading from the rim structure post track-making. The track topology is consistent with a two stage process in which the substrate first supports the heel, then

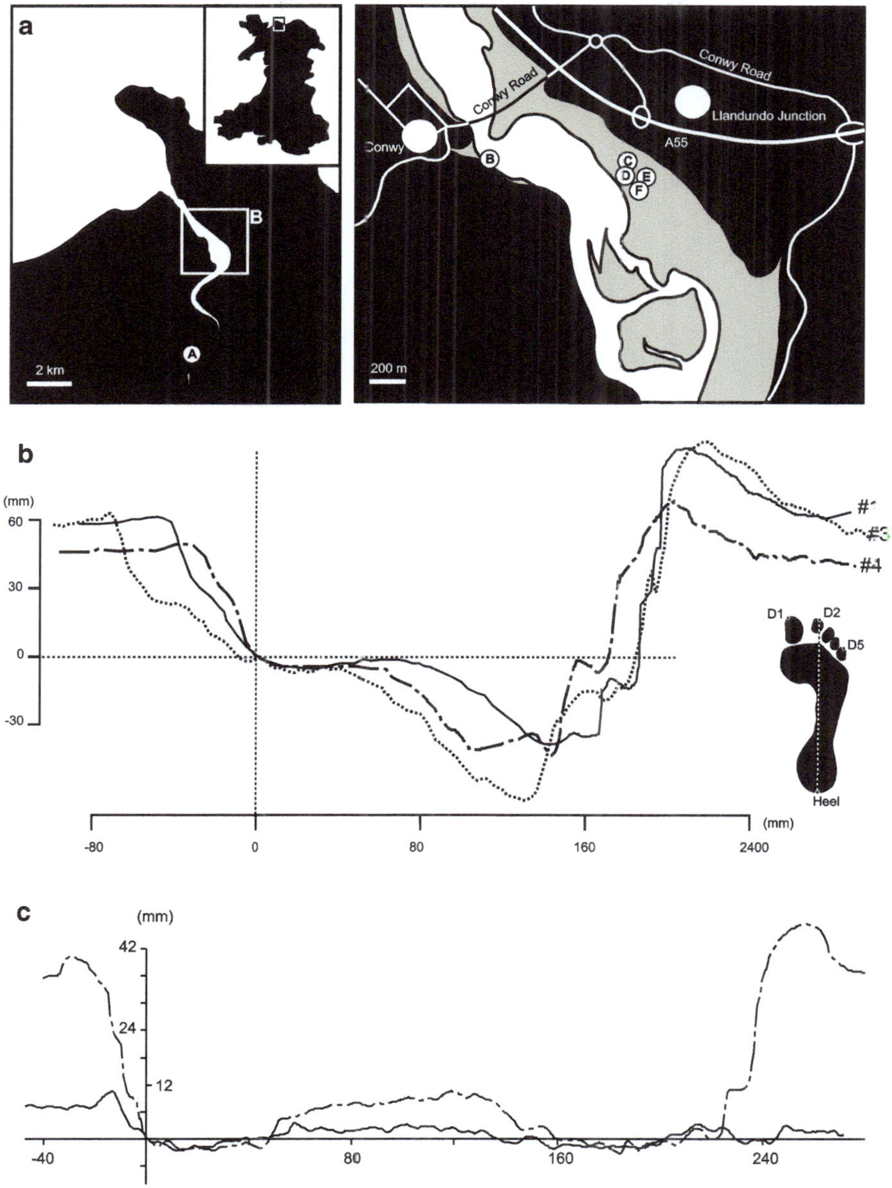

Fig. 5.6 (**a**) Location map of modern track sites around the margins of the Conwy Estuary in North Wales. (**b**) Longitudinal cross-sections through typical tracks at Site A on the Conwy Estuary. The origin has been set at the base of the heel to ensure that that the profiles are comparable. Note the wedge-shaped form. (**c**) Longitudinal cross-sections through typical tracks at Site B on the Conwy Estuary. The origin has been again set at the base of the heel to ensure that that the profiles are comparable. Note the fact that the deeper track has a more prominent medial longitudinal arch due to the availability of sediment to mould it

Fig. 5.7 Tracks at Site A on the Conwy Estuary (Fig. 5.6). Note the abducted first toe and the rim structure and the bulging track wall due to loading from the rim structure

fails again as additional pressure is applied during the latter part of stance (Fig. 5.6b). This is perhaps exacerbated by the need for the track-maker to reaccelerate foot motion in the later part of stance to counter the cushioning effect of the soft substrate during heel strike. This may cause the track-maker to lever the foot forward more energetically applying greater force. There may also be an element here in which the track-maker limits the heel contact themselves in light of the substrate's lack of strength.

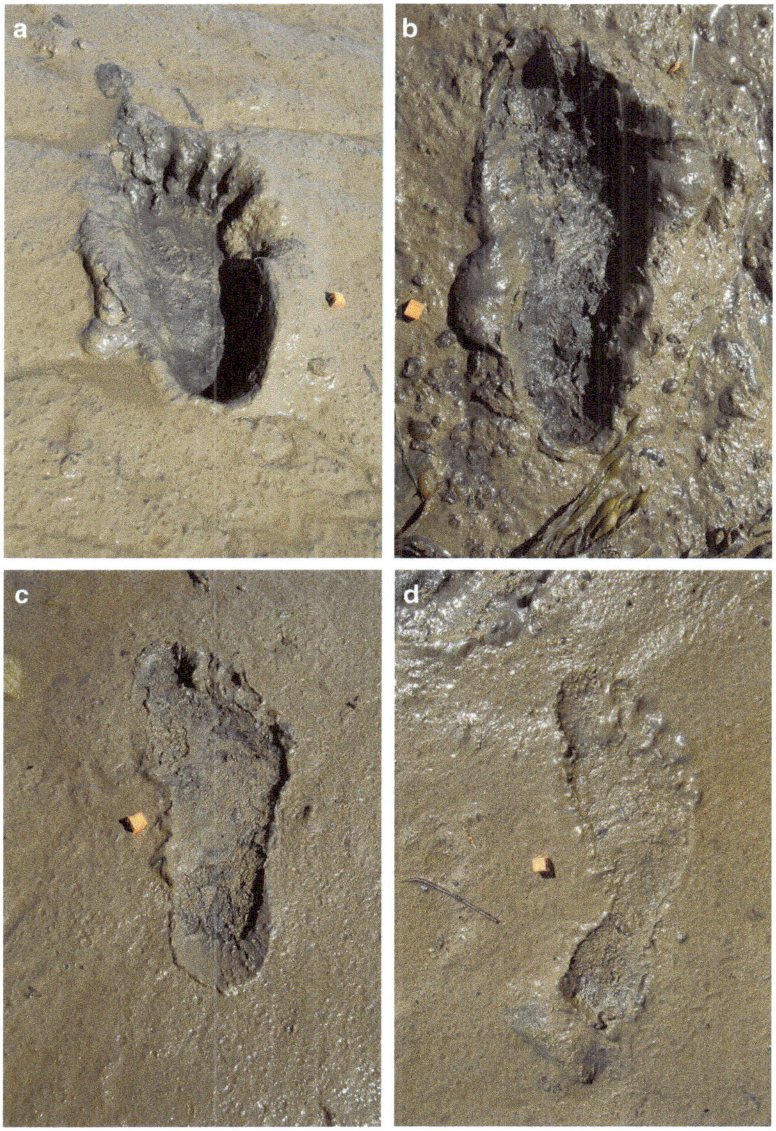

Fig. 5.8 Typical tracks at Site B on the Conwy Estuary (Fig. 5.6). The overall topology is influenced by the thickness of the surface layer of mud-rich and saturated sediment

In contrast Site B (Fig. 5.6a) contains a firm sub-layer of sand and granule gravel (29 % = water) below a surface layer of mud (42 % = silt + clay; 78 % = water) of varying thickness (<5 to 150 mm). The topology of individual tracks within the trackway is primarily a function of the thickness of the surface layer (Fig. 5.8); where the mud layer was thickest the tracks have a flat base with little surface relief and prominent rim-structures reflecting the lateral displacement of the mud

(Fig. 5.8a, b). In contrast where the firm sub-base outcrops at the surface the tracks are shallow with only those areas of maximum plantar pressure being recorded (Fig. 5.8c, d). In practice as illustrated by the longitudinal cross-sections in Fig. 5.6c there is little difference in the morphology of the plantar surface irrespective of the presence or absence of the mud layer, although the presence of the mud is seen to mould the medial longitudinal arch more effectively. It is worth noting that the overall topologies of the tracks – plantar surface + track walls + rims – are different however (Fig. 5.8). The third site (Site C; 66 % = silt + clay; 51 % = water; Fig. 5.6a) shows a similar picture in which track topology was controlled by the thickness and moisture content of a soft mud-rich surface layer overlying the firmer sandier substrate (Figs. 5.9 and 5.10). The rim structures form narrow ridges around the margins of the track in keeping with the theoretical observations of Allen (1997; Fig. 5.5a). Shallower tracks show evidence of distal sediment movement in thin slabs between the track base and the sole of the foot (Fig. 5.10). Sites D to G (Fig. 5.6a) are all similar in terms of their properties with higher sand content (78–91 %) and moisture levels (29–34 %), although some have firmer sub-layers. Track depth is very variable between these sites ranging from 20 mm to as much as 60 mm. Longitudinal asymmetry in track depth – heel to ball/toe depth – is more prominent in deeper tracks where there is no firm sub-layer, although this is absent where a firm sub-layer is encountered. A few tracks recovered from Site E show a different type of sub-layer effect; here a sand layer overlies a compact, consolidated silt layer. This gives the tracks a very ragged outline caused by sand adhesion on the foot during extraction separating thin sheets of sand from the underlying silts.

These observations show as one would expect that mud and moisture content of a substrate are important in determining track topology along with the nature and presence of the vertical variations in sedimentary properties. Sites B and C illustrate the importance of a near surface sub-layer, while Sites D to G have similar grain-sizes and moisture contents but track depth varies strongly being deepest where the sediment has the lowest level of natural consolidations and therefore bulk density. If one assumes crudely that track depth correlates with the bearing capacity of the substrate it is possible to use depth as a proxy for substrate strength. On this basis one can pool the data and examine track topology at different depths. Figure 5.11 is based on a series of longitudinal profiles extracted from tracks at various sites illustrative of a range of substrate bearing capacities. In all these cases the sediment is homogenous with no near-surface sub-layer present. The weakest substrate, represented by the deepest tracks, is illustrated by Profile-6; it is triangular in shape with the plantar surface increasing in angle in a distal direction starting with a sloping or inclined heel wall, passing into a gently inclined heel and mid-foot before steepening forward of the mid-foot. Profile-5 is similar but here the heel and mid-foot area is much more horizontal. The weight appears to be first borne by the substrate, only to fail as the pressure is shifted forward in the latter part of stance as acceleration occurs. As the substrate becomes firmer the longitudinal asymmetry in depth is maintained but becomes gradually less pronounced such that the longitudinal medial arch is more pronounced (Profile-4) and increasingly emphasised by the proximal movement of sediment

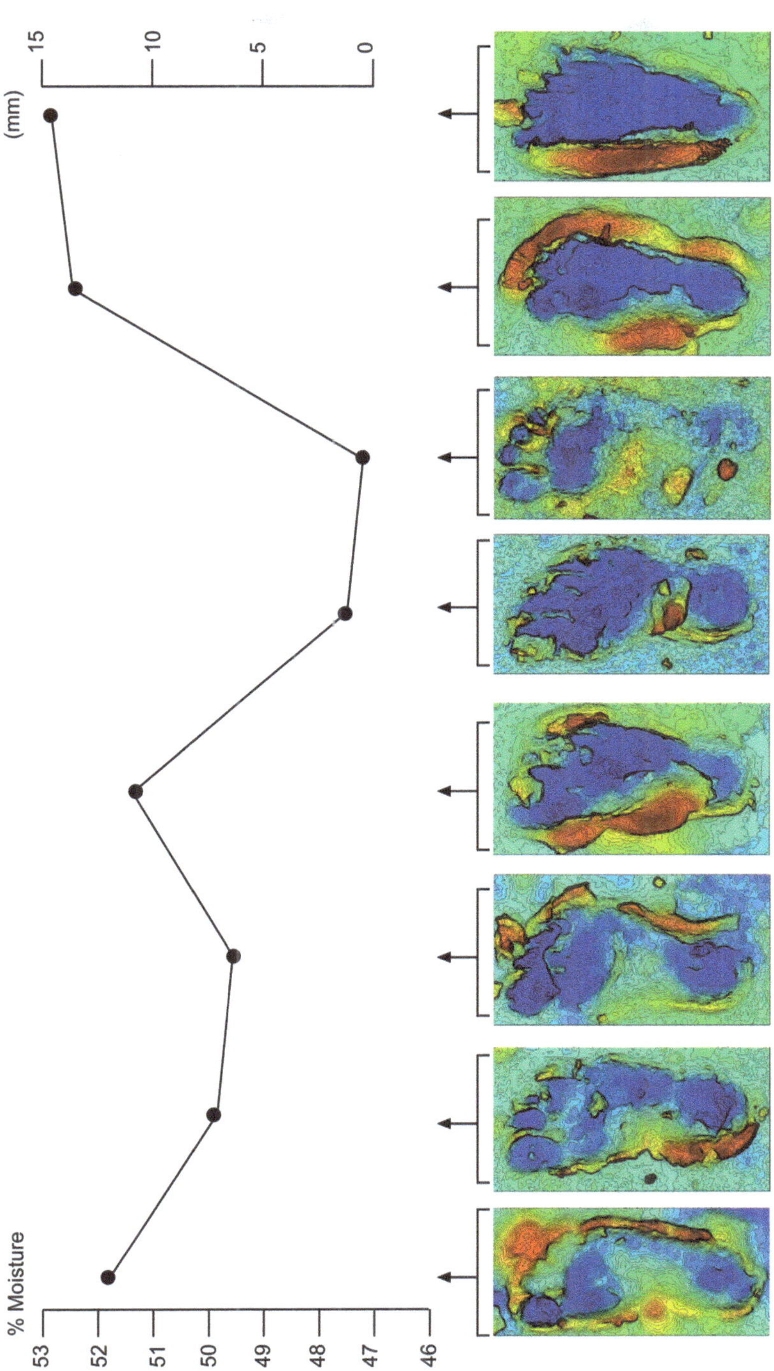

Fig. 5.9 Tracks from Site C Conwy Estuary (Fig. 5.6). Showing again the influence on track topology of the thickness and moisture content of the surface layer

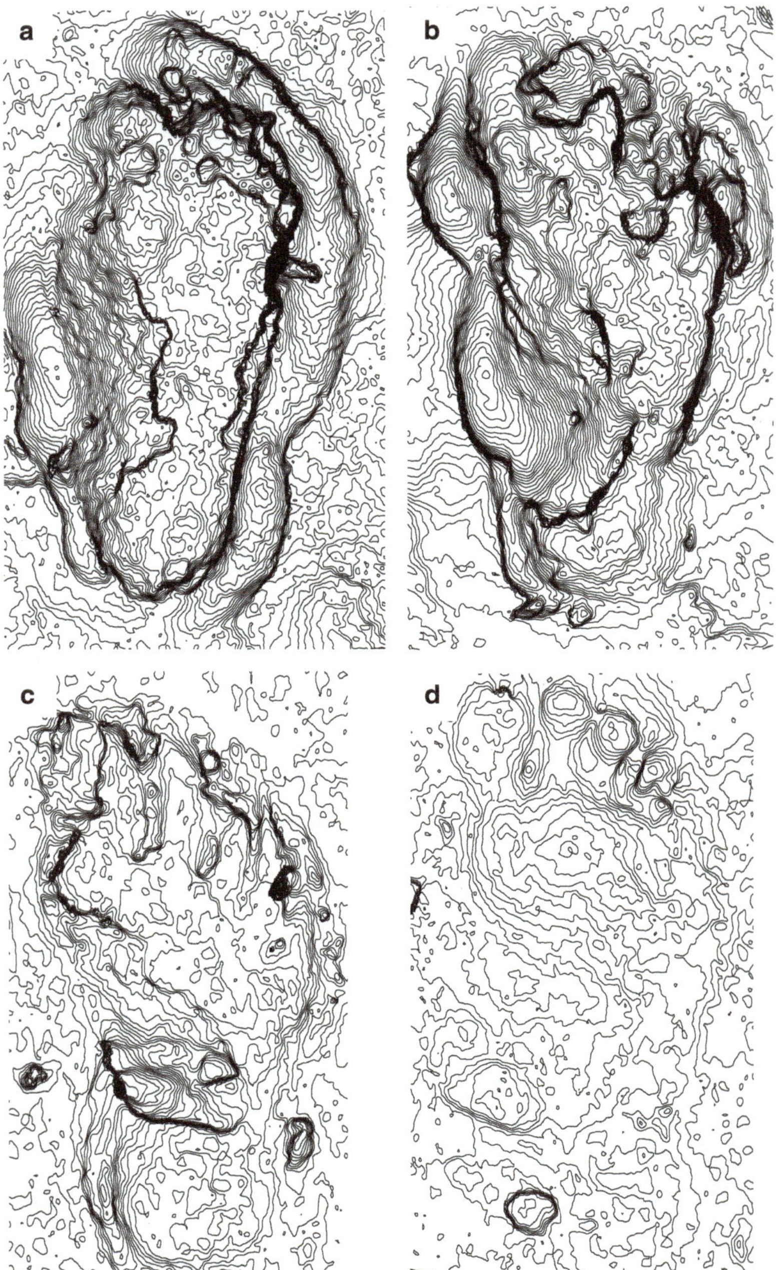

Fig. 5.10 Contour maps for typical tracks at Site C on the Conwy Estuary (Fig. 5.6). Contour interval is 1 mm. The overall topology is influenced again by the thickness of the surface layer of mud-rich and saturated sediment

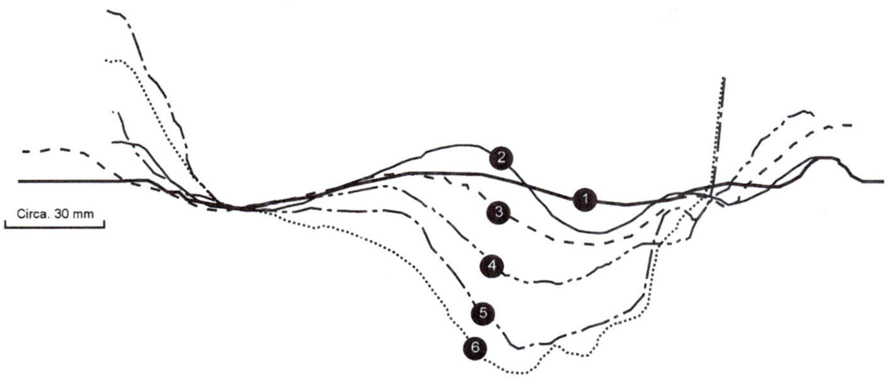

Fig. 5.11 Typical cross-sections from a selection of the Conwy Estuary tracks showing how longitudinal track form varies with substrate strength

under rotation beneath the ball of the foot. As the substrate becomes firmer still the depth asymmetry heel to ball becomes even less pronounced; the degree to which the medial longitudinal arch is enhanced by sediment movement accounts for much of the variation. The degree of medial weight transfer seems to be critical here accounting for the differences between Profile-2 (more marked medial transfer) and Profile-3 for example. Finally in Profile-1 only those areas of maximum plantar pressure are recorded by the track and there is typically more equity in depth between the heel and ball areas, with a slight tendency for the heel to be deeper. One has to be careful not to over-infer from a limited data set of this sort, but Fig. 5.11 provides a working hypothesis of how track topology may vary with substrate strength emphasising the importance of overall sediment strength rather than any one given variable such as the mud-sand ratio or moisture content for example. This model only applies in the absence of a firm sub-layer which has a profound influence on the track topology whatever the substrate properties.

It is possible to explore this further using both Harry's and Harriet's trackways from Walvis Bay (see Sects. 1.4 and 3.2.3). The tracks here were made in flat lying sheets of silt and fine sand deposited by seasonal flood waters constrained by and escaping between large dunes. These two sub-parallel trackways were examined by Morse et al. (2013). The longest trackway (Harry's) of over 70 tracks has length of at least 54 m starting at a shallow runnel in the south. A uniform step (0.65 ± 0.03 m) and stride length (1.38 ± 0.02 m) throughout the trackway was used by Morse et al. (2013) to argue for a consistent and steady pace on the part of the track-maker. Parts of the trackway where underlain by in situ, firmer sediment and here well-developed rim structures where present, as well as evidence in the tracks for both longitudinal slippage and internal/external rotation consistent with the individuals' feet failing to gain the traction necessary to maintain forward motion before and during the mid-stance and final stages of stance. At the northern end of the trackway the surrounding areas shows an increase in animal trampling which has destroyed any sublayers present producing a more homogenous substrate. Here the trackway becomes progressively deeper with individual tracks taking on a more trapezoidal shape in planform. This typological variability causes basic track dimensions to vary (Figs. 2.13 and 5.12). For example, track length, defined as the distance from the proximal heel edge to the distal tip of the second toe, varied by as much as ±60 mm associated primarily with: (1) extension of the toes due to forward drag; (2) internal and external rotation of the foot as the individual effectively lost traction and 'skated' on the mud; and (3) longitudinal track compression due to proximal slippage in the later part of stance. Both heel and ball width show less variability, although both decrease as print depth increases at the northern end of the trail (Fig. 5.12). Morse et al. (2013) were able to divide the trackway into four sections each representing a different type of substrate condition. They calculated a mean track for each section using Pedobarographic Statistical Parametric Mapping (pSPM; see Sect. 2.7) and related the topological variations to substrate as follows (Fig. 5.12):

1. Runnel. The substrate here was assumed to have been weaker at the time of imprinting due to the higher water content in the runnel which drained flood waters from the area. The tracks show a much higher degree of depth variation than elsewhere in the trackway with some tracks having a strong longitudinal depth asymmetry (enhanced ball/hallux depths) while others don't. There is an absence of displacement rims, implying that strain was accommodated primarily through compression. It would appear that in some cases the substrate bore the subjects' weight forming a track of similar depth, while in others cases it compressed to a greater degree under the ball/toe areas during the latter phases of stance thereby giving a stepped longitudinal profile.
2. Firm. In this section of the trackway the grain-size is finer, with undisturbed primary bedding and a lower salt content suggesting a firmer drier substrate during imprinting. Longitudinal asymmetry in track depth is less pronounced with the deepest points occurring more equitably in both the heel and toe/ball region. The medial longitudinal arch is well-defined as are individual toe pads, and marginal displacement rims are common suggesting that strain accommodation occurred via both sediment displacement and compression. There is evidence of plantar slippage and foot rotation within the near-surface layer which appears to have acted as shallow shear zone between a more stable sub-layer and the plantar surface of the foot. Track morphology owes much to the undisturbed sublayer in this section of the trackway.
3. Lightly trampled. In this area, the track-maker overprinted animal tracks, the substrate grain-size remains fine and the salt content remains high. The animal tracks are significant here in disturbing the natural stratigraphy and sublayer, as well as elevating the sediment moisture content by retaining pockets of water. The tracks are similar to those in the previous section from which they grade in topology gradually deepening and developing deeper ball regions as the level of animal trampling increases. The medial longitudinal arch of the tracks is enhanced by the movement of sediment below the ball of the foot and in some cases the arch merges with prominent rim structures. The rim structures are

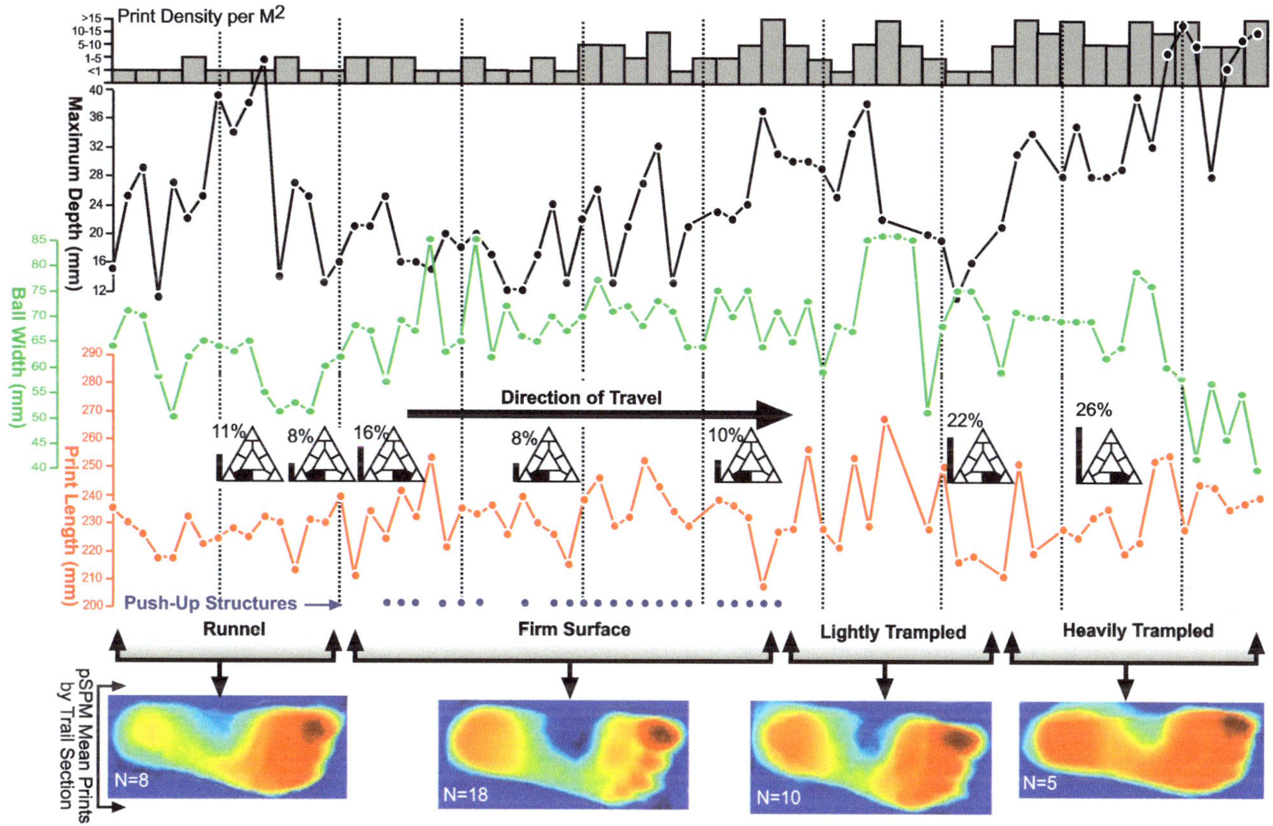

Fig. 5.12 Summary data for Harry's Trackway in Namibia. The trackway has been divided up into lengths on the basis of substrate and mean tracks calculated for each section (see Sect. 1.4) (Modified from Morse et al. (2013))

much broader in width than in the previous section reflecting a thicker mobile surface layer of sediment.

4. Heavily trampled. Toward the end of the trackway the density of surrounding animal prints increases markedly as does the salt content of the sediment and track depth all suggesting a decrease in the original bearing capacity of the substrate (Fig. 5.12). Overall, as the track depth increases the variation in depth between the heel, ball/toe areas decreases and the plantar surface becomes more uniform and flatter reflecting a firmer sub-layer at depth (i.e. below the depth of trampling). The tracks have narrower heels and a strong trapezoidal form when viewed from above. Displacement rims are absent which suggests that strain was accommodated largely by compression.

These observations indicate that small-scale variations in substrate properties, particularly water content, at the time of imprinting and animal trampling influence track topology. Deeper prints are found in softer substrates, and appear to have more subdued medial longitudinal arches. Intermediate strength substrates are associated with movement of sediment proximal to the ball medially, enhancing the medial longitudinal arch. Increased longitudinal depth asymmetry (i.e. deeper ball than heel) is associated with substrates of more variable strength. The tracks collectively do not give therefore a reliable indication of the morphology of the longitudinal medial arch. The presence of clear vertical stratigraphy in substrate properties is critical to the observed variations. Prominent, narrow rim structures occur where a firmer sub-layer was close to the surface and the shallower tracks record more of the plantar detail of the foot and are less affected by the movement of sediment beneath the ball.

The second trackway (Harriet) present at the Walvis Bay site was also explored by Morse et al. (2013) and its analysis has to reinforce these patterns. This trackway is shorter, consisting of only 18 tracks and also has a consistent stride (0.976 ± 0.09 m) and step length (0.37 ± 0.01 m). The trackway traversed the runnel and is associated with slightly higher clay content and is assumed to have had higher water content at the time of imprinting. There are two distinct track topologies related to depth (Fig. 5.13). The shallow topology is associated with marginal areas of the runnel and slightly coarser grain sizes. These tracks consist of a heel strike or contact zone, poorly-defined ball and prominent hallux. The tracks lack displacement rims and in some cases the heel impression is almost absent and only areas of maximum plantar pressure are recorded in the prints. As the substrate appears to gain strength this becomes increasingly restricted to the hallux and ball area alone. The deeper track topology has a well-defined heel, ball and toe area with proximal shear beneath the foot in selected tracks.

Both Harry's and Harriet's trackways were used by Morse et al. (2013) to develop a schematic model of how track topology varies with substrate (Fig. 5.14). The observations are similar to those we made at the start of this section from the Conwy Estuary and the two models Figs. 5.11 and 5.14 form natural complements. It is worth emphasising that one of the most significant controls on substrate strength in the Namibian case according to Morse et al. (2013) is the level of animal trampling

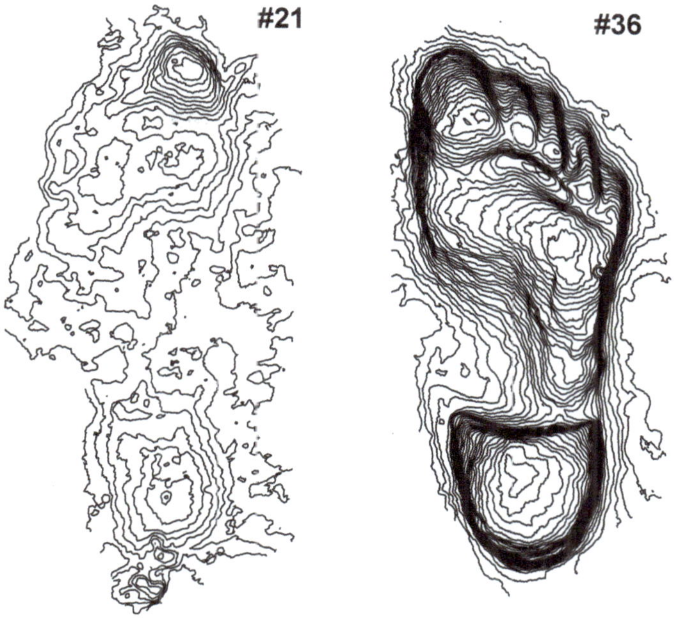

Fig. 5.13 Typical track topologies for Harriet's Trackway in Namibia (see Sect. 1.4)

prior to track-making and its role in homogenising the substrate, removing vertical stratigraphic variations and promoting water retention within the sediment through puddling in the tracks.

The presence of prominent rim structures is a feature of the tracks in the Namibia example especially where a firmer sub-layer was present near the surface. They are also a particularly striking feature of the Acahualinca tracks site in Nicaragua (Schmincke et al. 2010; see Sect. 3.2.1). These tracks are formed in a tephra layer 30–150 mm thick. Individual tracks vary significantly in depth with some tracks having prominent rim structures while adjacent tracks don't. Schmincke et al. (2010) suggest that this may reflect a combination of: variations in the thickness of the ash layer over the underlying more competent substrate below; differences in the load applied by the track-makers due to such things as age, sex and load carrying; as well as the result of individuals walking in each other's tracks.

Bates et al. (2013) modelled the influence of a firmer sub-layer on track topology and also the fundamental assumption that track depth links in some direct way to dynamic foot pressure which in turn is linked to the biomechanics of the track-maker's limbs. They used a combination of modelling and modern beach tracks combined with laboratory pressure records for the same subjects to explore this idea. In their numerical simulations plantar pressure was found to exceed depth under the

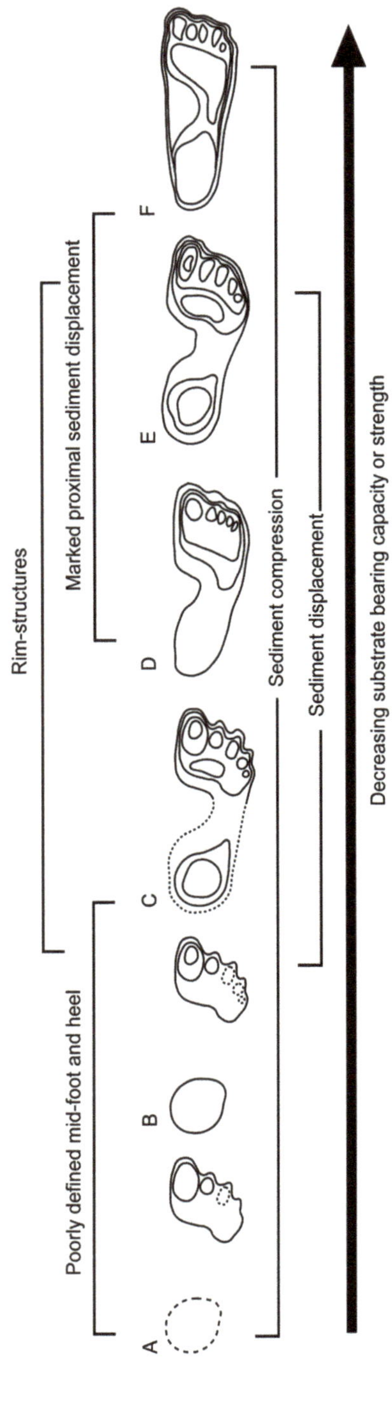

Fig. 5.14 Model of track topology and substrate (Modified from Morse et al. (2013))

forefoot, but the converse was true at the heel, especially when a firm sub-layer was introduced. This relationship held over a range of depths but became more pronounced. For example, in deep track simulations, without a mechanically firm sub-layer, relative pressure exceeded depth across the entire track, except under the heel. Understanding the consolidation and ultimately the resistance of the substrate to dynamic loading after initial failure is critical to explaining these pressure-depth relationship and Bates et al. (2013) suggest that discrete regions of the plantar foot surface have different potential to indent the substrate. We have already seen that Falkingham et al. (2010) demonstrated that the ratio between surface area and edge length of an indenter influenced the degree of substrate penetration with shapes that have relatively less edge length for a given area penetrating more for a given pressure than shapes with more intricate outlines. In the case of a human track the sub-circular heel has a low edge to area ratio compared to the forefoot and the toes. Bates et al. (2013) suggest that this may explain the contrast in pressure-depth relationships they observed between the heel and forefoot. With the introduction of a firm sub-layer the areas in which pressure exceed depth become more restricted to the forefoot, while under the heel the disparity increased with depth greater than pressure. They also suggested that introducing a sub-layer decreased overall track depth by the order of 16 % (Bates et al. 2013) and that pressures increased under the forefoot since rapid compression under the heel enables the foot to lever forward more effectively exerting enhanced forefoot pressures. This result ties in with the observations of Morse et al. (2013) which suggests that on the firmest substrates only the ball and first toe are visible and the heel outline may be repressed. Looking at their beach track data Bates et al. (2013) also noted differences between the spatial distribution of pressure in a record and depth in a track, for example heel impressions of tracks differ from the pressure records, which are more elongated longitudinally, in that sediment is displaced medially, laterally and in some cases posteriorly to produce a more shortened and rounded impression compared to the pressure record. Deeper beach tracks were characterised by relatively greater forefoot depths that exceeded pressure, reversing the relationship found in shallower prints in which depth exceeded pressure only in the heel area. This may reflect the deceleration caused by softer substrates which cushions the heel and requires the track-maker to re-accelerate during the latter parts of stance through the application of greater plantar pressure. We have already seen how deeper tracks often show greater depth asymmetry longitudinally. The work of Bates et al. (2013) demonstrates that at a first order there is a relationship between the distribution of plantar pressure and track depth, but that this varies across different areas of the track and crucially is depth dependent with the correlation – pressure to depth – being closest for shallower tracks. Substrate strength – a crucial control on track depth – is therefore critical in determining the pressure-depth relationship and therefore any biomechanical inferences that can be made from a given sequence of tracks. The idea that shallow tracks record the best level of biomechanical detail is an interesting one, since in many respects they have the least preservation potential (Allen 1997). A site that is notable for

Fig. 5.15 Influence of microbial mats. (**a**) Close up of a mat on a surface of sediment, note the surface salt crystals. (**b**) A track from which the surface layer of sediment bound by a microbial mat has been rolled back. (**c**) Taponomic effect of microbial mats. The sequence of images shows the build-up of microbial mats following track formation, while they may obscure the topology of the track they form a protective layer and aid sedimentation. All these images are kindly provided by Daniel Marty and reproduced here with permission

preserving such shallow impressions is that at Cuatro Ciénegas in northern Mexico (Fig. 3.11; Gonzalez et al. 2009; Felstead et al. 2014). Here the tracks are preserved in tufa around the margins of groundwater fed pools (Pozas) and the surface appears to have been particularly non-compliant with a thin surface layer of damp carbonate-rich mud to record the prints. Only those areas with maximum plantar pressure are recorded.

This emerging picture of the control that substrate has on track topology is summarised in such models as Figs. 5.11 and 5.14, but is complicated by the fact that natural sediment once exposed is usually colonised by bacteria and algae moderating natural substrate properties (Fig. 5.15a, b; Thulborn 1990). This idea was explored in a seminal paper by Marty et al. (2009) and while their focus was inter-tidal environments their work is relevant to all trackways. Microbial mats grow on any damp sediment especially in abundant sunlight and warm conditions. In particular Marty et al. (2009) show that they are ubiquitous on modern (and ancient) carbonate and siliciclastic tidal flats (Gerdes and Krumbein 1994;

Schieber 2007). Benthic microbial communities either agglutinate sediment particles to sticky surfaces or act as baffles that trap sediment. In addition they may enhance lithification through the precipitation of carbonates (Dupraz and Visscher 2005). Marty et al. (2009) explored the influence of such microbial mats on human tracks and in particular the modification that they cause to substrate properties. Effectively microbial mats form a surface layer which has its own properties – yield strength and elasticity – whose behaviour will impact on the sediment below and therefore the preservation of a track within it. In the case of dry consolidated mats overlying dry unconsolidated sediment, tracks will only form if the track-maker exerts sufficient pressure to crack the mat and penetrate the sediment below (Fig. 5.16a; Marty et al. 2009). A thicker mat may help maintain sediment moisture below allowing deeper tracks to form (Fig. 5.16b, c). The best tracks are formed with thicker moist mats with firm but deformable sediment beneath (Fig. 5.16d). Effectively the surface mat acts as another variable with its own moisture and therefore elastic properties through which the interaction between the foot and the ground is moderated. In general the quality of the tracks decrease as the water content of both the mat and the under lying sediment increase and the best preservation conditions are favoured by intermediate sediment- and mat-moisture conditions (Marty et al. 2009). It is worth noting that mats may help sediment deformation and rim structures are common where the mat is thick plastic, moist to water-unsaturated and sitting on top of moist to water unsaturated sediment (Fig. 5.2b, c).

5.4 Track Taphonomy

Taphonomy is the branch of palaeontology that deals with the processes of fossilization, in this context the modification of a track from the point a track-maker's foot is withdrawn to the point at which the track is buried and/or lithified and by extension to the point at which it is exhumed and exposed either naturally or via excavation. Clearly understanding how a track is modified post-formation is critical to ensuring that any inferences made from it are based on its true topology and not some artefact caused by its preservation. We have already seen how taphonomy starts with the immediate withdrawal of the foot (Fig. 5.1c, d) and Morse et al. (2013) illustrate the importance of track-wall suction as a foot is withdrawn.

There is another fundamental aspect here however and that is 'survivorship' a term used by Cohen et al. (1993) to describe the probability that a track will make it into the fossil record. In the context of Lake Manyara in Tanzania they emphasised the importance of both animal trampling and near-shore wetting and drying; the moist sediment of the strandline zone preserves the best tracks but is also a focus for animal trampling, daily wetting and drying as well as shoreline erosion. Laporte and Behrensmeyer (1980) argued that preservation of a track involved a subtle trade-off between the rate of trampling and the rate of burial (Fig. 5.17). Scott et al. (2010) demonstrated through both field observations and laboratory simulations that clay

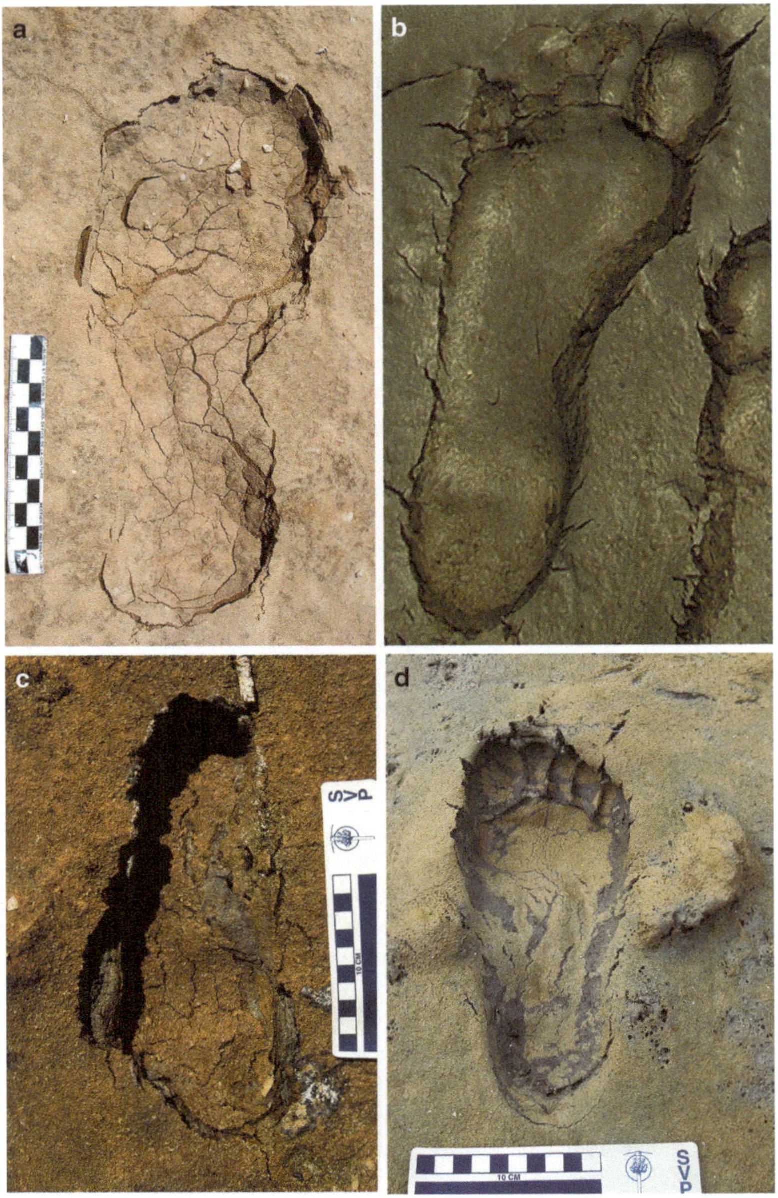

Fig. 5.16 Influence of microbial mats. (**a**) Dry mat lying above dry sediment. (**b**) Wet mat lying above damp sediment. (**c**) Thicker mat above firm sediment. (**d**) Moist mat above moist sediment producing ideal conditions for track formation and preservation (All these images are kindly provided by Daniel Marty and reproduced here with permission)

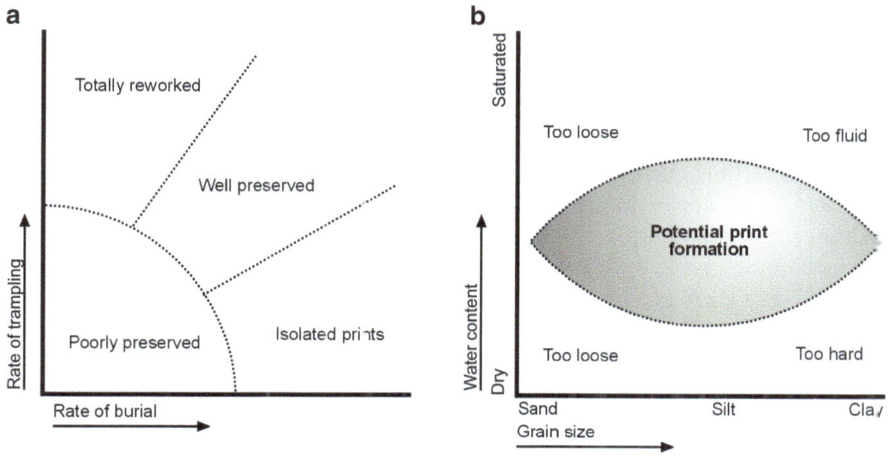

Fig. 5.17 Environmental and sedimentological variables associated with the formation and preservation of human tracks (Modified from Laporte and Behrensmeyer (1980))

mineralogy impacts on the volume changes and desiccation cracking regime that occurs as tracks are subject to wetting and drying. The nature of salt growth within the sediment is controlled by such things as the groundwater level and composition; chemistry and lake level fluctuations; and rainfall regimes. Cohen et al. (1993) show that salt growth hinders the preservation of smaller tracks. Scott et al. (2008) provides a detailed record of the preservation history of a given tracked surface on the Sandai Plain (Lake Bogoria, Kenya) which emphasised the importance first of stabilisation by desiccation, soil-crusting and the growth of organic films before cementation occurred via calcite, analcime and zeolites, along with the weathering of minor clay minerals and iron-magnesium oxides.

Exposure to the elements and to the processes of weathering and erosion all play an important role in determining the final topology of a preserved track and whether it is a complete true track, some relict part, or an under track (Henderson 2006; Table 1.1). Allen (1997) points out that tracks may be subject to erosion either before burial as well as post-exposure or in theory as part of a complex exhumation and burial history. Shallow tracks may record the best anatomical detail (Bates et al. 2013) but the deeper ones have the better preservation potential. Marty et al. (2009) has shown how the growth of microbial mats may play an important role in track preservation, since while the growth of mats may obscure the visible form of a track (Fig. 5.15c) they act to protect and to trap sediment causing rapid infilling to form over-tracks and may enhance lithification via the precipitation of carbonates. McKee (1947) demonstrated the role sediment adhesion to dew lining a track as a potential infill process for tracks. This process has also been observed by the authors in the Namibian desert where coastal fogs dampens the sediment surface with a potent mix of water and sea salt to which wind-blown sands sticks like glue! The rate at which a surface desiccates and the presence of pore-water or groundwater salts is likely to be critical especially to the precipitation of potential cements (Scott et al. 2010). The surface relief, proximity of the water table and continued surface wetting either via rainfall or by some form of over-wash are all important.

The seasonal rainfall filled water hole shown in Fig. 5.18 from Amboseli in Kenya provides some illustration of the variables at play. The margin of the small wetland area illustrated is extremely desiccated forming a non-compliant surface that encroaches slowly inwards as the environment dries out between rainfall events. Continued animal trampling in this area not only destroys any existing tracks and consolidates the surface, but also creates a surface breccia of silt blocks and delaminated flakes that moves in advance of the desiccated margin infilling tracks. The surface topography of the wetter areas is irregular and pitted by large tracks including those of elephants and giraffe as well as by a range of smaller antelopes, zebra and birds. The area shows multiple generations of desiccation cracks. Microbial mats occur across the surface in a variety of different forms and in areas of standing water tracks can be seen to be filled by turbid silts from which suspension settling occurs. All the tracks show some degree of modification and long-term preservation is likely to be limited unless an extreme widespread episode of sustained flooding leads to deposition of a thick capping layer of silt. Complete desiccation might also lead to modified preservation, but it is hard to see how subsequent animal trampling and wind erosion would not just destroy the surface. Observations like this stress the importance to long-term preservation by rapid burial following track formation. This needs to be achieved by passive widespread deposition rather than by a concentrated high energy and therefore potentially erosive flow.

Scott et al. (2010) provides a summary of the key variables in the formation and taphonomy of animal tracks around Lake Bogoria in Kenya drawing on the work of others and is reproduced here in a modified and abridged form:

1. Frequency of lake inundation, proximity to the shoreline and substrate moisture content in general (Mckee 1947; Van der Lingen and Andrews 1969; Tucker and Burchette 1977; Frey and Pemberton 1986; Lockley 1986; Lockley et al. 1987; Scrivner and Bottjer 1986; Cohen et al. 1993; Sadler 1993; Brand 1996; Allen 1997; Avanzini et al. 1997; Zhang et al. 1998; Gatesy et al. 1999; Fornós et al. 2002; Ashley and Liutkus 2003; Manning 2004; Uchman and Pervesler 2006; Platt and Hasiotis 2006; Milàn 2006; Davis and Stevenson 2007; Scott 2007, Scott et al. 2008, 2010; Buatois and Mángano 2009; Marty et al. 2009; Jackson et al. 2009; Genise et al. 2009). Lake level fluctuations will control the exposure time of a tracked surface and its renewal via erosion or by burial under fresh sediment. Lake shore proximity has also been found to be important in determining substrate saturation and therefore its ability to mould the track-maker's feet. In general the water content of sediment and its history during a track's taphonomy is probably, as illustrated above, the most widely cited influence on track formation and taphonomy.
2. Depth of the water table and pore-water salinity (Lockley 1986; Loope 1986; Cohen et al. 1993; Ashley and Liutkus 2003; Manning 2004; Scott 2007, Scott et al. 2008, 2010; Marty et al. 2009). This will again control the substrate's

Fig. 5.18 Plate of photographs showing the taphonomic variable at play around a watering hole in the Amboseli National Park in Kenya

moisture content and also provide a reservoir for capillary rise feeding salt efflorescence which may impact on the preservation of tracks. A near-surface water table may also help support the growth of microbial mats. The chemistry of pore water is also important to the development and type of salt efflorescence present, the swelling behaviour of some clay minerals and to the potential for cementation all of which may influence the survivorship of a track.

3. Evaporation and desiccation regime (Tucker and Burchette 1977; Laporte and Behrensmeyer 1980; Scrivner and Bottjer 1986; Loope 1986; Frey and Pemberton 1986; Cohen et al. 1993; Avanzini et al. 1997; Zhang et al. 1998; Ashley and Liutkus 2003; Genise et al. 2009; Scott et al. 2010). This will also determine a substrate's moisture content as well as the formation of desiccation cracks and the growth of salt efflorescence or microbial mats. Evaporation may also lead to the formation of protective crusts composed of salt and/or clay. The development of desiccation cracks has a destructive potential, especially for small track. It also encourages the creation of surface relief which may retain moisture and facilitate the growth of microbial mats. In the Amboseli example shown in Fig. 5.18 desiccation is clearly associated with the formation of a surface breccia of silt lamina and blocks.
4. Surface hydrology (Tucker and Burchette 1977; Laporte and Behrensmeyer 1980; Scott et al. 2010). Rainfall events will affect the moisture content of sediment and ponding within track and may be important in water retention. Salt efflorescence will also be affected by rainfall and the dissolution of salt growths may reveal new substrates for track preservation. Sheet floods and ephemeral stream flow have both a destructive role and one that might lead to track burial as well as again changing the moisture content of surface sediments. The break-up of surface crusts and the transport of surface detritus including desiccation breccia may all be important. Ponding of water within small surface pools may also lead to the deposition of protective mud drapes and will impact on the cycle of desiccation and 'sun-baking'.
5. Clay mineralogy, substrate chemistry and organic content (Tucker and Burchette 1977; Laporte and Behrensmeyer 1980; Loope 1986; Scrivner and Bottjer 1986; Fornós et al. 2002; Davis and Stevenson 2007; Scott 2007, Scott et al. 2010). The absolute quantity of clay will influence the substrate's consistency and its behaviour under wetting and drying. Higher amounts of clay may retard the break-up of interstitial salts and therefore the formation of sediment crusts and add cohesive strength to a substrate. The exact type of clay present will have an important impact on the width and nature of the desiccation cracks that develop and also the swelling potential given subsequent additions of moisture all of which will impact on the survivorship of tracks. Organic content promotes substrate cohesiveness and assist in its subsequently stabilisation.
6. Presence of roots, microbial mats and biofilms (Laporte and Behrensmeyer 1980; Frey and Pemberton 1986; Scarboro and Tucker 1995; Avanzini et al. 1997; Kvale et al. 2001; Scott 2007, Scott et al. 2010; Genise et al. 2009; Marty

et al. 2009). The presence of roots adds strength to sediment and therefore aids track preservation, but they may also hinder the quality of preservation of finer anatomical details. The peat tracks described by Bennett et al. (2010), for example, contain little anatomical detail and demonstrate the challenge. The growth of live roots is an issue and may lead to the break-up of a surface. We have already discussed earlier in this chapter the potential of a microbial mat to influence substrate properties and track formation/preservation. Their role in stabilising tracks and assisting in their burial and/or cementation is clear and in particular they may combine with salt efflorescence to form protective crusts.

7. Bioturbation (Laporte and Behrensmeyer 1980; Scrivner and Bottjer 1986; Lockley 1986; Frey and Pemberton 1986; Allen 1997; Cohen et al. 1993; Allen 1997; Kvale et al. 2001; Fornós et al. 2002; Ashley and Liutkus 2003; Scott 2007, Scott et al. 2009, 2010; Genise et al. 2009; Morse et al. 2013). We have already seen how vertebrate trampling may alter substrate properties by destroying natural stratigraphy, homogenising sediment's consistency through mixing and leading to enhanced moisture content by aiding water retention (Morse et al. 2013). The role of trampling in relation to a track's potential survivorship has been widely documented and is clearly a key variable in the potential for a track to enter the fossil record. Trampling may also break-up surface crusts aiding surface deflation and promoting the creation of a desiccation breccia such as that seen in the Amboseli example (Fig. 5.18). Invertebrate bioturbation also has role in first homogenising sediment and then secondly it may play a part in disturbing a surface, although it may also enhance infiltration of cement rich waters.
8. Deflation (McKee 1947; Lewis and Titheridge 1978; Tucker and Burchette 1977; Laporte and Behrensmeyer 1980; Loope 1986; Frey and Pemberton 1986; Cohen et al. 1993; Fornós et al. 2002; Davis and Stevenson 2007; Scott et al. 2010). The action of the wind on a tracked surface is widely cited as a cause of track loss, not only via direct erosion but by the disturbance of salt and desiccation crusts. The deposition of windblown sand may help to promote burial especially where it adheres to damp surface around the margins of tracks.

The taphonomic processes associated with track formation have been summarised by Marty et al. (2009) and is reproduced in Fig. 5.19 in a modified form to include longer term geological events. This model identifies Formation > Modification > Preservation > Burial and Digenesis as the key steps. Most of the processes which facilitate disturbance have been summarised above and are mitigated by early cementation and/or rapid burial or as pointed out by Marty et al. (2009) by the growth of microbial mats that both protect and trap sediment assisting in the infill and cementation of a track. The model can be extended to include the long-term burial history of a series of tracks (Fig. 5.19), given that excessive compression and/or tectonic activity may impact on the long-term chances of track preservation. In particular the potential for tracks to be exhumed naturally, modified and then re-buried cannot be discounted in some environments.

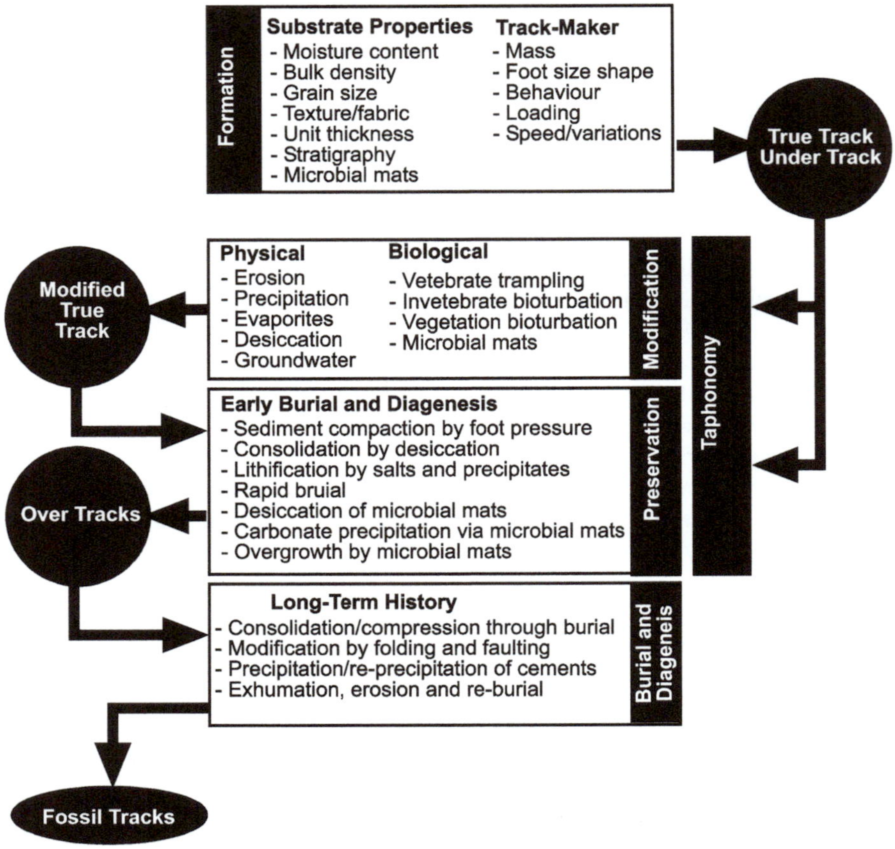

Fig. 5.19 Model of the variables in play during the fossilisation of a vertebrate track (Modified from Marty et al. (2009))

5.5 Summary

There is no doubt that substrate consistency plays an important role in determining the topology of tracks that result from the interaction of human feet with the ground. Modelling this is not straight forward since sediments are complex admixtures involving: inorganic grains and/or fragments; organic filaments and mats; along with various combinations of moisture and air. Track topology is not sensitive to variation in just one of these variables but to a combination of them all. Variations in moisture content, for example, are important but have to be seen in the context of other variables. Equally a clear distinction between fine-grained soils in the engineering sense (<μ63 m) and coarser sediments is also not clear cut. Instead track topology seems to vary in a fairly consistent way with a more general measure of substrate bearing capacity, moderated by the presence or absence of vertical variations in sediment properties due to natural stratigraphy. So in a uniform soft-substrate a deep track with a typically asymmetrical (heel to ball/toe depths) or wedge-shaped longitudinal profile tends to occur. The substrate first holds the weight of the individual and then fails further as the weight shifts forward during the latter stages of stance. This may be exacerbated further in part the impact of the substrate on the biomechanics of the track-maker and the cushioning effect of the soft-substrate on heel strike and the consequent need for the track-maker to re-accelerate gait in the latter stages thereby enhancing toe-off pressures and track depths. This longitudinal asymmetry slowly moderates as the substrate strengthens and the tracks shallow such that on firm substrates only those areas with the highest plantar pressures are preserved especially around the medial side of the ball and the first two toes (Figs. 5.11 and 5.14). This picture is complicated by the introduction of stratigraphy and the intendant vertical variation in sedimentary properties. The work of Bates' et al. (2013) demonstrates that depth does correlate in some way with biomechanical pressure but that it varies across different areas of the foot and that the ratio of area to length may be an important factor in determining the relative penetration of the heel and ball areas with respect to the applied pressures. They also significantly, show that the relationships are best in shallow tracks and tend to breakdown and/or complicate for deeper ones.

There are some real positives to be taken from the above since generic models of track topology and substrate do appear to hold across different sediment admixtures and therefore environments which is encouraging when trying to make comparisons between different sites. It is not the depositional environment that matters but more the generic comparison of sites with substrates of a similar bearing capacities and track depths. This picture is complicated by the introduction of surface layers in the form of microbial mats which are naturally prevalent in most track forming environments (Marty et al. 2009) and need to be factored into the interpretation of human tracksites. Post-formation the taphonomic processes which impact on the track topology that is ultimately fossilised are considerable (Fig. 5.19) and of these we emphasise the importance of algal growths and the hydrogeology of a particular site and therefore its desiccation history. Added to this, survivorship in the context of trampling by other track-makers, whether by human or animals, is critical.

References

Allen JRL (1989) Short paper: fossil vertebrate tracks and indenter mechanics. J Geol Soc 146(4):600–602

Allen JRL (1997) Subfossil mammalian tracks (Flandrian) in the Severn Estuary, SW Britain: mechanics of formation, preservation and distribution. Philos Trans R Soc Lond B352:481–518

Ashley G, Liutkus CM (2003) Tracks, trails and trampling by large vertebrates in a rift valley paleo-wetland, lowermost Bed II, Olduvai Gorge, Tanzania. Ichnos 9:23–32

Avanzini M, Frisia S, Van Den Driessche K et al (1997) A dinosaur tracksite in an early Liassic tidal flat in northern Italy; paleoenvironmental reconstruction from sedimentology and geochemistry. Palaios 12(6):538–551

Bates KT, Savage R, Pataky TC et al (2013) Does footprint depth correlate with foot motion and pressure? J R Soc Interface 10(83):20130009
Bennett MR, Harris JWK, Richmond BG et al (2009) Early hominin foot morphology based on 1.5 million year old footprints from Ileret, Kenya. Science 323:1197–1201
Bennett MR, Gonzalez S, Huddart D et al (2010) Probable neolithic footprints preserved in inter-tidal peat at Kenfig, South Wales (UK). Proc Geol Assoc 121:66–76
Bird RT (1944) Did Brontosaurs ever walk on land? Nat Hist 53:60–67
Brand LR (1996) Variations in salamander trackways resulting from substrate differences. J Paleontol 70:1004–1010
Bromley RG (2001) Tetrapod tracks deeply set in unsuitable substrates: recent musk oxen in fluid earth (East Greenland) and Pleistocene caprines in Aeolian sand (Mallorca). Bull Geol Soc Den 48:209–215
Brown T (1999) The science and art of tracking. Berkley Trade, New York
Buatois LA, Mángano MG (2009) Applications of ichnology in lacustrine sequence stratigraphy: potential and limitations. Palaeogeogr Palaeoclimatol Palaeoecol 272(3):127–142
Buckland W (1928) Note sur les traces de tortues observées dans les grès rouge. Ann de Sci Nat Paris 13:85–86
Cham R, Redfern MS (2002) Changes in gait when anticipating slippery floors. Gait Posture 15:159–171
Charteris J, Wall JC, Nottrodt JW (1981) Functional reconstruction of gait from the Pliocene hominid footprints at Laetoli, northern Tanzania. Nature 290:496–498
Charteris J, Wall JC, Nottrodt JW (1982) Pliocene hominid gait: new interpretations based on available footprint data from Laetoli. Am J Phys Anthropol 58:133–144
Cohen A, Lockley MG, Halfpenny J et al (1993) Modern vertebrate tracks from Lake Manyara, Tanzania and their paleobiological implications. Paleobiology 19:433–458
Craig RF (2004) Soil mechanics. E and FN Spon Chapman and Hall, London, UK
Crevier-Denoix N, Robin D, Pourcelot P et al (2010) Ground reaction force and kinematic analysis of limb loading on two different beach sand tracks in harness trotters. Equine Vet J 42(s38):544–551
Crompton RH, Pataky TC, Savage R et al (2012) Human-like external function of the foot, and fully upright gait, confirmed in the 3.66 million year old Laetoli hominin footprints by topographic statistics, experimental footprint-formation and computer simulation. J R Soc Interface 9:707–719
D'Aout K, Meert L, Van Gheluwe B et al (2010) Experimentally generated footprints in sand; analysis and consequences for the interpretation of fossil and forensic footprints. Am J Phys Anthropol 141:515–525
Davis BA, Stevenson AC (2007) The 8.2 ka event and Early–Mid Holocene forests, fires and flooding in the Central Ebro Desert, NE Spain. Quat Sci Rev 26:1695–1712
Demathieu GR (1987) Thickness of footprint reliefs and its significance: research on the distribution of the weights upon the autopodia. Glossary and manual on tetrapod footprint palaeoichnology. Dep Nac da Produçao Mineral, Brasilıa
Diedrich C (2002) Vertebrate track bed stratigraphy at new megatrack sites in the Upper Wellenkalk member and orbicularis member (Muschelkalk, Middle Triassic) in carbonate tidal flat environments of the western Germanic Basin. Palaeogeogr Palaeoclimatol Palaeoecol 183:185–208
Dupraz C, Visscher PT (2005) Microbial lithification in marine stromatolites and hypersaline mats. Trends Microbiol 13:429–438
Falkingham PL (2012) Acquisition of high resolution three-dimensional models using free, open-source, photogrammetric software. Palaeontol Electronica 15:1T:15p
Falkingham PL, Margetts L, Manning PL (2010) Fossil vertebrate tracks as paleopenetrometers: confounding effects of foot morphology. Palaios 25(6):356–360
Falkingham PL, Bates KT, Margetts L et al (2011) The 'Goldilocks' effect: preservation bias in vertebrate track assemblages. J R Soc Interface 8:1142–1154
Farlow JO (1989) Ostrich footprints and trackways: implications for dinosaur ichnology. Dinosaur tracks and traces. Cambridge University Press, Cambridge

Felstead NJ, Gonzalez S, Huddart D et al (2014) Holocene-aged human footprints from the Cuatrociénegas Basin, NE Mexico. J Archaeol Sci 42:250–259
Fornós JJ, Bromley RG, Clemmensen LB, Rodríguez-Perea A (2002) Tracks and trackways of *Myotragus balearicus* Bate (Artiodactyla, Caprinae) in Pleistocene aeolianites from Mallorca (Balearic Islands, Western Mediterranean). Palaeogeogr Palaeoclimatol Palaeoecol 180:277–313
Frey RW, Pemberton SG (1986) Vertebrate lebensspuren in intertidal and supratidal environments. Holocene barrier islands, Georgia. Senckenberg Marit 18:45–95
Gatesy SM, Middleton KM, Jenkins FA et al (1999) Three-dimensional preservation of foot movements in Triassic theropod dinosaurs. Nature 399:141–144
Genise JF, Melchor RN, Archangelsky M et al (2009) Application of neoichnological studies to behavioural and taphonomic interpretation of fossil bird-like tracks from lacustrine settings: the Late Triassic–Early Jurassic? Santo Domingo Formation, Argentina. Palaeogeogr Palaeoclimatol Palaeoecol 272:143–161
Gerdes G, Krumbein WE (1994) Peritidal potential stromatolites – a synopsis. In: Bertrand-Sarfati J, Monty C (eds) Phanerozoic stromatolites II. Springer, Dordrecht
Gonzalez AH, Lockley MG, Rojas CS, Espinoza JL, Gonzalez S (2009) Human tracks from Quaternary tufa deposits, Cuatro Ciénegas, Coahuila, Mexico. Ichnos 16:12–24
Henderson DM (2006) Simulated weathering of dinosaur tracks and the implications for their characterization. Can J Earth Sci 43:691–704
Jackson SJ, Whyte MA, Romano M (2009) Laboratory-controlled simulations of dinosaur footprints in sand: a key to understanding vertebrate track formation and preservation. Palaios 24:222–238
Kvale EP, Johnson AD, Mickelson DL et al (2001) Middle Jurassic (Bajocian and Bathonian) dinosaur megatracksites, Bighorn Basin, Wyoming, USA. Palaios 16:233–254
Laporte LF, Behrensmeyer AK (1980) Tracks and substrate reworking by terrestrial vertebrates in Quaternary sediments of Kenya. J Sediment Res 50:1337–1346
Lewis DW, Titheridge DG (1978) Small scale sedimentary structures resulting from foot impressions in dune sands. J Sediment Res 48:835–837
Lockley MG (1986) The paleobiological and paleoenvironmental importance of dinosaur footprints. Palaios 37–47
Lockley MG (1991) Tracking dinosaurs: a new look at an ancient world. Cambridge University Press, Cambridge
Lockley MG, Rindsberg AK, Zeiler RM (1987) The paleoenvironmental significance of the nearshore Curvolithus ichnofacies. Palaios 2:255–262
Lockley M, Roberts G, Kim JY (2008) In the footprints of our ancestors: an overview of the hominid track record. Ichnos 15:106–125
Loope DB (1986) Recognizing and utilizing vertebrate tracks in cross section: Cenozoic hoofprints from Nebraska. Palaios 1:141–151
Manning PL (2004) A new approach to the analysis and interpretation of tracks: examples from the Dinosauria. In: McIlroy D (ed) The application of ichnology to palaeoenvironmental and stratigraphical analysis, vol 228, Geol Soc, London, Special Publ., pp 93–128
Marty D, Strasser A, Meyer CA (2009) Formation and taphonomy of human footprints in microbial mats of present-day tidal-flat environments: implications for the study of fossil footprints. Ichnos 16(1–2):127–142
McKee ED (1947) Experiments on the development of tracks in fine cross-bedded sand. J Sediment Res 17:23–28
Melchor RN, de Valais S, Genise JF (2002) Bird-like fossil footprints from the Late Triassic. Nature 417:936–937
Melchor RN, Bedatou E, de Valais S et al (2006) Lithofacies distribution of invertebrate and vertebrate trace-fossil assemblages in an Early Mesozoic ephemeral fluvio-lacustrine system from Argentina: implications for the Scoyenia ichnofacies. Palaeogeogr Palaeoclimatol Palaeoecol 239:253–285
Meldrum DJ (2004) Fossilized Hawaiian footprints compared with Laetoli hominid footprints. In: Meldrum DJ, Hilton CE (eds) From biped to strider. Springer, New York, pp 63–83
Meldrum DJ, Lockley MG, Lucas SG et al (2011) Ichnotaxonomy of the Laetoli trackways: the earliest hominin footprints. J Afr Earth Sci 60:1–12

Milàn J (2006) Variations in the morphology of emu (*Dromaius novaehollandiae*) tracks reflecting differences in walking pattern and substrate consistency: ichnotaxonomic implications. Palaeontology 49:405–420

Milàn J, Bromley RG (2008) The impact of sediment consistency on track and undertrack morphology: experiments with Emu tracks in layers cement. Ichnos 15:18–24

Morse SA, Bennett MR, Liutkus-Pierce C et al (2013) Holocene footprints in Namibia: the influence of substrate on footprint variability. Am J Phys Anthropol 151:265–279

Padian K, Olsen PE (1984) The fossil trackway Pteraichnus: not pterosaurian, but crocodilian. J Paleontol 58:178–184

Platt BF, Hasiotis ST (2006) Newly discovered sauropod dinosaur tracks with skin and foot-pad impressions from the Upper Jurassic Morrison Formation, Bighorn Basin, Wyoming, USA. Palaios 21:249–261

Sadler CJ (1993) Arthropod trace fossils from the Permian De Chelly Sandstone, northeastern Arizona. J Paleontol 67:240–249

Scarboro DD, Tucker ME (1995) Amphibian footprints from the mid-Carboniferous of Northumberland, England: Sedimentological context, preservation and significance. Palaeogeogr Palaeoclimatol Palaeoecol 113:335–349

Schieber J (2007) Microbial mats on muddy substrates–examples of possible sedimentary features and underlying processes. Atlas of Microbial Mat Features Preserved Within the Siliciclastic Rock Record. Atlas Geosci 2:117–133

Schmincke HU, Rausch J, Kutterolf S, Freundt A (2010) Walking through volcanic mud: the 2,100 year-old Acahualinca footprints (Nicaragua) II: the Acahualinca people, environmental conditions and motivation. Int J Earth Sci 99:279–292

Scott JJ (2007) Biogenic activity, trace formation, and trace taphonomy in the marginal sediments of saline, alkaline, Lake Bogoria, Kenya Rift Valley. In: Bromley RG, Buatois LA, Mágano MG, Genise J (eds) Sediment-organism interactions: a multifaceted ichnology, vol 88, SEPM, Special Publication., pp 309–330

Scott JJ, Renaut RW, Owen RB (2008) Preservation and paleoenvironmental significance of a footprinted surface on the Sandai Plain, Lake Bogoria, Kenya Rift Valley. Ichnos 15:208–231

Scott JJ, Renaut RW, Buatois LA et al (2009) Biogenic structures in exhumed surfaces around saline lakes: an example from Lake Bogoria, Kenya Rift Valley. Palaeogeogr Palaeoclimatol Palaeoecol 272:176–198

Scott JJ, Renaut RW, Owen RB (2010) Taphonomic controls on animal tracks at Saline, alkaline Lake Bogoria, Kenya Rift Valley: impact of salt efflorescence and clay mineralogy. J Sediment Res 80:639–665

Scrivner PJ, Bottjer DJ (1986) Neogene avian and mammalian tracks from Death Valley National Monument, California: their context, classification and preservation. Palaeogeogr Palaeoclimatol Palaeoecol 57:285–331

Thulborn T (1990) Dinosaur tracks, vol 410. Chapman and Hall, London

Tucker ME, Burchette TP (1977) Triassic dinosaur footprints from South Wales: their context and preservation. Palaeogeogr Palaeoclimatol Palaeoecol 22:195–208

Tuttle R (1990) The pitted pattern of Laetoli feet. Nat Hist 99:60–65

Uchman A, Pervesler P (2006) Surface lebensspuren produced by amphipods and isopods (crustaceans) from the Isonzo delta tidal flat, Italy. Palaios 21:384–390

Van der Lingen GJ, Andrews PB (1969) Hoof-print structures in beach sand: NOTES. J Sediment Res 39:350–357

White TD, Suwa G (1987) Hominid footprints at Laetoli: facts and interpretations. Am J Phys Anthropol 72:485–514

Zhang G, Buatois LA, Mángano MG et al (1998) Sedimentary facies and environmental ichnology of a Permian playa-lake complex in western Argentina. Palaeogeogr Palaeoclimatol Palaeoecol 138:221–243

Chapter 6
Inferences from Human Tracks

Abstract What can, and perhaps more importantly cannot, be inferred from a series of human tracks? In this chapter we explore this question by first looking at the relationships that exist between various foot dimensions and such things as stature and body mass. We explore the population specific nature of these empirical relationships and demonstrate their limitations with respect to the interpretation of tracks in the geological record. In the latter part of the chapter we explore what can deduced about the speed of a track-maker and the way in which variations in that speed may be reflected in the topology of the tracks themselves.

6.1 The Limits of Inference

Having looked at the influence of substrate on track topology in the previous chapter it is appropriate now for us to focus on the inferences that can, and perhaps more importantly cannot be made from a series of human tracks. The literature on which one can draw from to make these inferences is varied, derived from a wide cross-section of disciplines including forensic science and anthropology. The geoarchaeologist is faced with a bewildering range of sources with which to interpret an ancient human trackway and must select from them carefully and with caution. There are unfortunately examples within the human track literature where the enthusiasm of the authors has exceeded the limitation of their data and the empirical models used to interpret it. Some have claimed, for example, to know the sex of a track-maker and their medical history all of which as we will see are not easily inferred from a track alone. Tuttle (2008) provides a cautionary read and is critical of the inferences that have been made in some forensic cases in the past. If you were to read the popular, and in many ways charming book, on tracking by Brown (1999) however you would be forgiven for believing that a trained tracker can deduce almost everything about a track-maker short of their national insurance number! There is no doubt that experienced trackers can follow a trackway and say something about the track-maker, but it is important to be cautious and clear about the assumptions that are being made at all times.

The foot changes and develops with age up until adult maturity and potentially these changes allow one to make inferences about a track-maker's age. Equally, empirical

M.R. Bennett, S.A. Morse, *Human Footprints: Fossilised Locomotion?*,
DOI 10.1007/978-3-319-08572-2_6

studies have shown that in a typical population of humans the length of various body parts are in proportion such that it is for example possible to infer something about the stature of a track-maker from the length of their tracks. It is important to note however that these are empirical formulations based on the statistical range that exists within a sample and this may, or may not, be representative of the population from which it is drawn and equally of the population from which the track-maker comes. All of these things are open to potential challenge. There are also relationships which link a track's dimensions to such things as body mass. So we have a broad range of potential inferences around body proportions that may be made about a track-maker from their tracks. Coupled to this is information on the spacing and the depth of tracks in a trackway which has the potential to provide inferences about the biomechanics of a track-maker's gait and the speed at which they were walking or running. We have therefore two broad categories of inference to explore, the first around body dimensions and the second around gait and locomotion.

6.2 Inferring Body Dimensions

The most frequently inferred dimension is stature, reflecting the fact that our feet exhibit a proportionate relation with total body height and for that matter with other body parts such as the hands (Wilder and Wentworth 1918; Kanchan et al. 2010a). Toppinard (1877) in a classic French work on physical anthropology proposed that a subject's foot length was between 14 % and 16 % of their height depending on the population from which they were drawn. This has given rise to the often quoted figure of 15 % which was broadly confirmed (14.9–18.1 %) by Barker and Scheuer (1998) and in several subsequent studies (e.g., Jasuja et al. 1991; Pawar and Pawar 2012) and was used in early studies of human tracks found in European caves (Vallois 1931; Pales 1976). The forensic potential of this was recognised by Macdonell (1902) and in the last 20 years a significant body of literature including numerous multipliers and regression equations with which to inferred stature from foot length have been published driven by interest from anthropologists and forensic scientists alike. The forensic focus is driven in part by the potential of feet, which often survive disasters and atrocities encased in footwear, to help with victim identification as well as criminal cases given that in large parts of the developing World habitual barefoot walking is still common place and unshod (naked) foot traces are frequently found at crime scenes (see Sect. 7.2). This literature provides an important resource, but not without limitations for the geoarchaeologist and as we have already said must be navigated with care.

Clearly empirical relationships only hold for the populations on which they are based and are dependent on the degree to which any given sample reflects the entire variation within that population (Atamturk and Duyar 2008). The fact that many samples focus on a limited age range, typically 20–25 years old and are often drawn from student populations may limit the degree to which they represent anything other than a group of young students! Equally other studies have drawn heavily on military populations (Robbins 1986; Adams and Herrmann 2009) and while many of these studies involve large samples one has to acknowledge that they may not be representative of the wider population since the military tend to select key physical attributes. Variation between populations due to race, ethnicity, nutrition and socio-economics will also be a factor limiting any analysis to the population from which it is drawn (Hrdlička 1935; Singh and Phookan 1993; Ashizawa et al. 1997; Kanchan et al. 2010b; Krishan et al. 2011b). For example, Sen and Ghosh (2008) developed regression models to infer height from foot length based on populations of 350 Rajbanshi and 100 Meche peoples from the Darjeeling district of West Bengal (India) and found that the two models were not cross-applicable without enhanced errors. It is also important to acknowledge, especially in dealing with ancient populations, that climate may also drive geographical variations in body proportions within both extant and extinct hominin populations (e.g., Roberts 1953; Katzmarzyk and Leonard 1998; Ruff 2002). One of the reasons that the literature has become so vast is the fact that each population potentially requires its own empirical model. There is also another complicating factor at work here which is that these studies are not consistently executed with respect to method; some use direct foot measurements (anthropometric), others two-dimensional prints made by either a static or dynamic track-maker and all use subtly different measures of basic foot dimensions as discussed in Sect. 2.6. In the context of fossil tracks few empirical relationship are based on three-dimensional data, with a couple of notable exceptions (Bennett et al. 2009; Dingwall et al. 2013). It is however worth pulling out a few pertinent points from this literature.

Many of the studies take length measurements from the most posterior point on the heel (Pternion) to the tip of each toe. Reel et al. (2010, 2012) noted that the correlations between stature and foot length were more significant for length measures between the fourth or fifth toes, a conclusion supported by some other studies (e.g., Fawzy and Kamal 2010), although not all (e.g., Ukoha et al. 2013). Any measure to the first toe must pass through the medial longitudinal arch which is a more flexible structure with multiple articulations in comparison to the lateral longitudinal arch which is more stable and therefore not subject to variations caused by tendon laxity, body mass or genetics (Dowling et al. 2001; Saltzman et al. 1995). As such Reel et al. (2012) argued that length measures to the lesser toes are more consistent measures of stature. The problem here is that the fifth toe pad is often absent from two-dimensional foot impressions, something which may reflect sex and ethnicity (Kulthanan et al. 2004; Reel et al. 2012). For example, Nataraja Moorthy et al. (2014) compared the non-occurrence of a 'fifth pad' in his study of Indian Tamils (5.5 % non-occurrence) to that found in other regional studies (Kanchan et al. 2012, 8 % Indian students; Nataraja Moorthy et al. 2011, 8.8 % Malaysian; Reel et al. 2012, 16.1 % UK) showing potential racial/ethnic variation. Reel et al. (2012) suggested that the non-occurrence of a 'fifth pad' was more common in males than females and hinted at the potential for this to help discriminate between the sexes. Using the Bournemouth data (see Sect. 1.4) 24 % of all males had a missing 'fifth pad' in static two-dimensional footprint trials which was just slightly higher than in the female sample (22 %). According to Reel et al. (2012) this may

Table 6.1 Comparison of estimated stature with actual stature for ten randomly selected individuals using regression models for each measure of foot length (heel to tip of each toe) for a population of Indian Tamils undertaken by Nataraja Moorthy et al. (2014)

Actual	PLT1	PLT2	PLT3	PLT4	PLT5	PRT1	PRT2	PRT3	PRT4	PRT5	Min	Max	Range	Mean
172	0.13	-0.47	-0.67	-1.01	-1.54	0.34	0.34	-0.45	-0.81	-1.39	-1.54	0.34	-1.88	171.4
173	0.12	-0.48	-0.99	-0.94	-0.96	0.51	0.05	-0.06	0.68	0.92	-0.99	0.92	-1.91	172.87
174	0.11	-0.26	-0.2	0.41	0.78	1.65	0.6	0.33	0.1	0.86	-0.26	1.65	-1.91	174.5
175	-1.09	-1.06	-0.59	-0.8	-0.48	-0.91	1.29	0.52	0.14	0.81	-1.09	1.29	-2.38	174.76
176.5	-1.6	-0.35	0.18	-0.37	-0.94	-1.24	0.31	-0.38	-0.95	-1.64	-1.64	0.31	-1.95	175.74
177	-0.31	-0.45	-1.14	-1.3	-0.45	-0.77	0.24	-1.27	-0.61	-0.96	-1.3	0.24	-1.54	176.2
178	0.08	-0.64	0.53	1.14	-0.32	-0.8	-1.23	0.11	0.26	-0.33	-1.23	1.14	-2.37	177.87
179	-1.52	-1.04	-1.5	-2	-1.54	-2.19	0.57	-1.88	-2.41	-2.25	-2.41	0.57	-2.98	177.25
180	-2.1	-2.24	-2.7	-2.58	-2.3	-1.2	1.09	-2.29	-2.78	-2.78	-2.78	1.09	-3.87	177.71
181	-1	-1.41	-1.44	-1.86	-2.87	-2.82	0.2	-1.7	-2.54	-3.3	-3.3	0.2	-3.5	179

Here we show the residuals – larger residuals picked out by the darker shading – calculated from Table 6 in Nataraja Moorthy et al. (2014). *PLT1* left print length to toe 1, *PRT1* right print length to toe 1 and so on

reflect differences in the pelvic tilt between males and females which consequently leads to differences in the centre of gravity and ultimately to the level flexion in the tendons which control the lesser toes. In the Nataraja Moorthy et al. (2014) study of over a thousand Indian Tamils they calculated the errors associated the different length measures for ten randomly selected prints (Table 6.1) and this gives an insight into the levels of variance involved with different regression models using different length measures, as well as the overall variance associated with stature estimates of this sort. Most studies also find a correlation between stature and foot breadth as well (e.g., Saxena 1984; Krishan 2008a, b; Zeybek et al. 2008; Reel et al. 2012) although this is usually much weaker and multiple regression models using several foot measures have been used by some workers (e.g., Giles and Vallandigham 1991; Gordon and Buiskstra 1992; Ozden et al. 2005; Sen and Ghosh 2008).

There are varying results with respect to the presence of bilateral asymmetry and therefore whether correlations should be based on mixed populations of right and left feet or just one (Table 6.1). In most cases the left foot is found to be larger than the right one although this is not universally true and the degree of asymmetry varies between the sexes (e.g., Manna et al. 2001; Tyagi et al. 2004; Agnihotri et al. 2007; Krishan 2008a, b; Sen and Ghosh 2008; Fawzy and Kamal 2010; Krishan et al. 2011a; Hairunnisa and Nataraja Moorthy 2014; Nataraja Moorthy et al. 2014). In contrast other studies have not found any marked asymmetry (Robbins 1985, 1986; Jasuja et al. 1991; Krishan and Sharma 2007; Zeybek et al. 2008; Hairunnisa and Nataraja Moorthy 2014). Bilateral asymmetry is independent of right-foot or right-hand dominance (Mysorekar et al. 1982) and its association with the left is believed to be linked to the fact that the left foot is typically more weight bearing and tends therefore to be associated with thicker bone development (Chhibber and Singh 1970; Rao and Kotian 1990). There is a slight suggestion in the range of studies that perhaps those with a more active (e.g., agricultural) habit may have greater bilateral asymmetry than those that don't but this has yet to be explored or validated. Bilateral

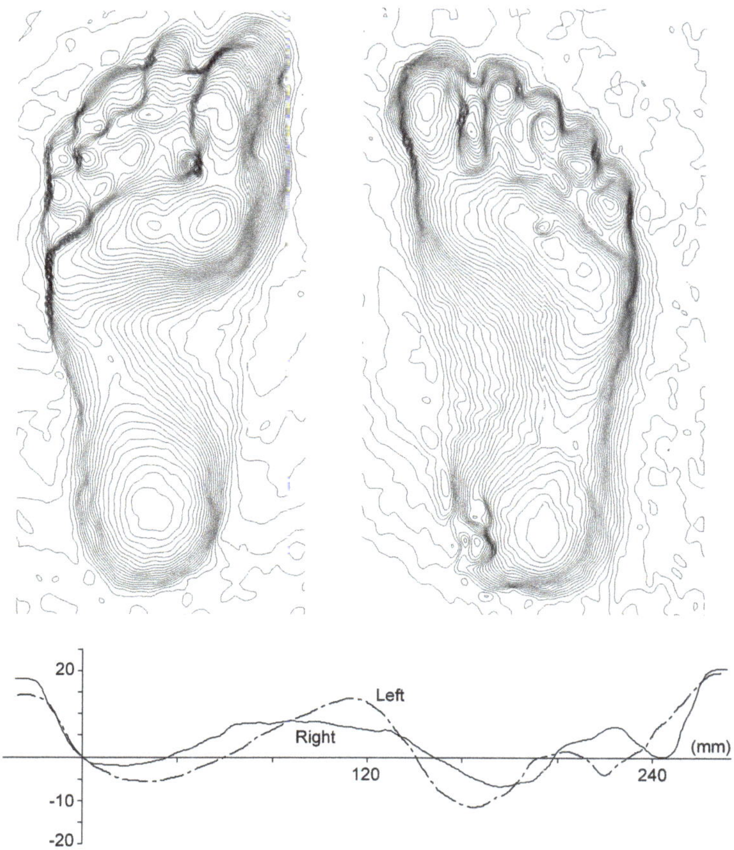

Fig. 6.1 Comparison of right and left tracks made in a combination of mud and sand for an individual showing slightly different foot functions. These are recurrent over multiple tracks. The longitudinal cross-sections clearly show the enhanced medial longitudinal arch caused by the greater medial weight transfer associated with the left foot

asymmetry may exist in other ways than simply with respect to length. Compare the right-left tracks in Fig. 6.1 which are similar in size and made by a subject with no known foot pathologies. These tracks show different patterns of foot function with the left foot showing evidence of a stronger medial weight transfer than the right.

These studies collectively show varying degrees of sexual dimorphism (Atamturk 2010; Hemy et al. 2013b), with female feet being generally smaller and in some cases narrower than male feet and as a consequence many authors have produced separate regression equations for both men and women, but the correlations are greater for pooled data in most cases (e.g., Nataraja Moorthy et al. 2013a, b; Hairunnisa and Nataraja Moorthy 2014) and sex is never known when dealing with fossil or unidentified tracks. The differences between the sexes is often more sophisticated than simply a question of size for example Hairunnisa and Nataraja Moorthy

(2014) in a study of Ibans adults from eastern Malaysia found that the males' feet tended to have longer first than second toes, a trait replicated in other Malaysian populations but that this difference was not pronounced in female feet (Nataraja Moorthy et al. 2013a, b). The differences are sufficient for a number of studies to attempt with good results to discriminate between the sexes statistically developing models from a multiple foot measurements (e.g., Zeybek et al. 2008; Atamturk 2010; Krishan et al. 2011a; Sen et al. 2011; Jowaheer and Agnihotri 2011; Hemy et al. 2013a; Uhrová et al. 2013). These are again population specific and while the models give good discrimination of cases typically between 80 and 90 %, it is hard to see how these could be applied to fossil tracks however unless the track-maker is certain to be drawn from a population which a discriminate model exists. Moudgil et al. (2008) uses the foot index (length/breadth; foot index) to explore sex differences, finding foot length and breadth values for males to be significantly higher than for females, but did not find a statistically significant relationship despite previous studies that suggested that one might exist (Tyagi et al. 2004).

A paper by Kinahan (2013) provides an interesting geoarchaeological perspective on this subject. He used not only different methods but also complementary lines of evidence to estimate stature of the Pre-Colonial human population of the Namib Desert (see Sect. 3.2.3). Firstly he used skeletal remains to estimate stature from both males and females following two different approaches (Wilson and Lundy 1994). Secondly, he estimated stature from fossil tracks using trackway averages and both the classic 15 % multiplier of Toppinard (1877) and the regression model of Kanchan et al. (2008) which was based on a sample of Gujjars Indians. Finally, he estimated stature from the grinding hollows made by kneeling individuals. The hollow provided an estimate of arm length allowing stature to be inferred accordingly (Bassey 1986; Mohanty et al. 2001). On the basis of the skeletal estimates strong sexual dimorphism was found with females being 89–92 mm shorter than males depending on the method used. In terms of the track inferences; the 15 % multiplier appeared to under estimate stature but the regression equation based on the Indian population gave values that were not statistically significantly different from those derived for the female skeletons. Grinding hollow estimates were smaller and more variable but again matched closely the female skeletal estimates. Kinahan (2013) argues that this reflects differential population sampling with the tracks and grinding hollows preferentially sampling females. The problem is that, as he acknowledges his sample of tracks may contain juveniles which might skew the results, but more importantly the foot-stature regression model based on an Indian population may not give the best size estimate for the Pre-Colonial population of the Namib. The study does however demonstrate nicely the power of converging and corroborative lines of evidence in making stature inferences. It also raises the important point that tracks may not sample a 'whole population' but that they may be preferentially left by those doing specific tasks, in the case of the Namib tracks watering goat/sheep flocks at the edge of flood waters. This activity based sampling has rarely been considered within human track studies.

The issue of whether sex can be determined from a fossil track site is an important one. A number of publications claim to have records of female track-makers and this is asserted with some conviction but little evidence in some cases (e.g., Roberts et al. 1996; Schmincke et al. 2010). As we have seen above, the forensic literature contains references to sexual dimorphism and to the identification of sex on the basis of discrimination models using multiple foot measurements for a given population. The problem is that these are difficult to apply in the fossil record not just because they are population specific but there are in reality a number of potential explanations for foot length variations including age, sex and natural population variability as well as differential preservation due to substrate effects. Simply identifying size differences is not enough to justify the recolonization of female track-makers. It is also worth pointing out that the recognition of specific pathologies and/or conditions (e.g., diabetic feet, pregnancy) has been claimed by some footprint observers (e.g., Roberts et al. 1996), but in truth there is no real data to support such inferences. Tuttle (1987) discusses a range of possible pathological explanations for the bilateral asymmetry in foot angles of the G-1 Laetoli trackway and within the animal track literature there are examples where 'limping' has been argued for (Ishigaki and Lockley 2010).

In comparison to the work on stature there are relatively few studies which examine the relationship between foot dimensions and body mass (Robbins 1985, 1986; Atamturk and Duyar 2008; Krishan 2008c) Statistically significant relationships have been noted although they are generally much weaker than with stature and are typically stronger with measures of foot width rather than length (Atamturk and Duyar 2008). It is worth noting that in the fossil footprint literature some workers have disputed the ability to infer mass from tracks (e.g., Tuttle et al. 1990). Krishan (2008c) explored this further by having his subjects carry additional weights (5 and 20 kg). The addition of the smaller weight had little impact but the larger weight was visible in the recorded foot dimensions enhancing the strength of the regression models accordingly.

None of the studies discussed so far actually deal with real three-dimensional tracks, at best they deal with two-dimensional pressure induced ink tracks or involve direct foot measurements. For those wishing to interpret three-dimensional tracks we need to explore the potential differences that may exist between these approaches. We can use the Bournemouth data introduced in Sect. 1.4 to have an initial look at this issue and more generally at the relationships that exist between tracks and the body dimensions of the track-maker. In this pilot analysis 19 landmarks were placed on contour maps of each track within Foot Processor. These landmarks were placed predominantly around the external margins of the track, using the first right and left tracks in a trackway unless obscured by deformation and/or ejecta in which case the second tracks were used (Fig. 6.2). In addition the two-dimensional tracks, made on pressure sensitive paper (see Sect. 1.4), were also analysed using a sub-set of identical landmarks with measurements being taken using a ruler and a protractor. Selected measurements are shown in Table 5.2.

First focusing on methods, the two-dimensional tracks made by standing subjects yield significantly different measures of foot dimensions than those obtained from the three-dimensional tracks made in the sand tray by walking subjects. Typically the two-dimensional data are between 10 and 20 mm smaller. If you compare foot length (H-D2) the differences range from as little as 2.7 to a massive 41.7 mm with a mean

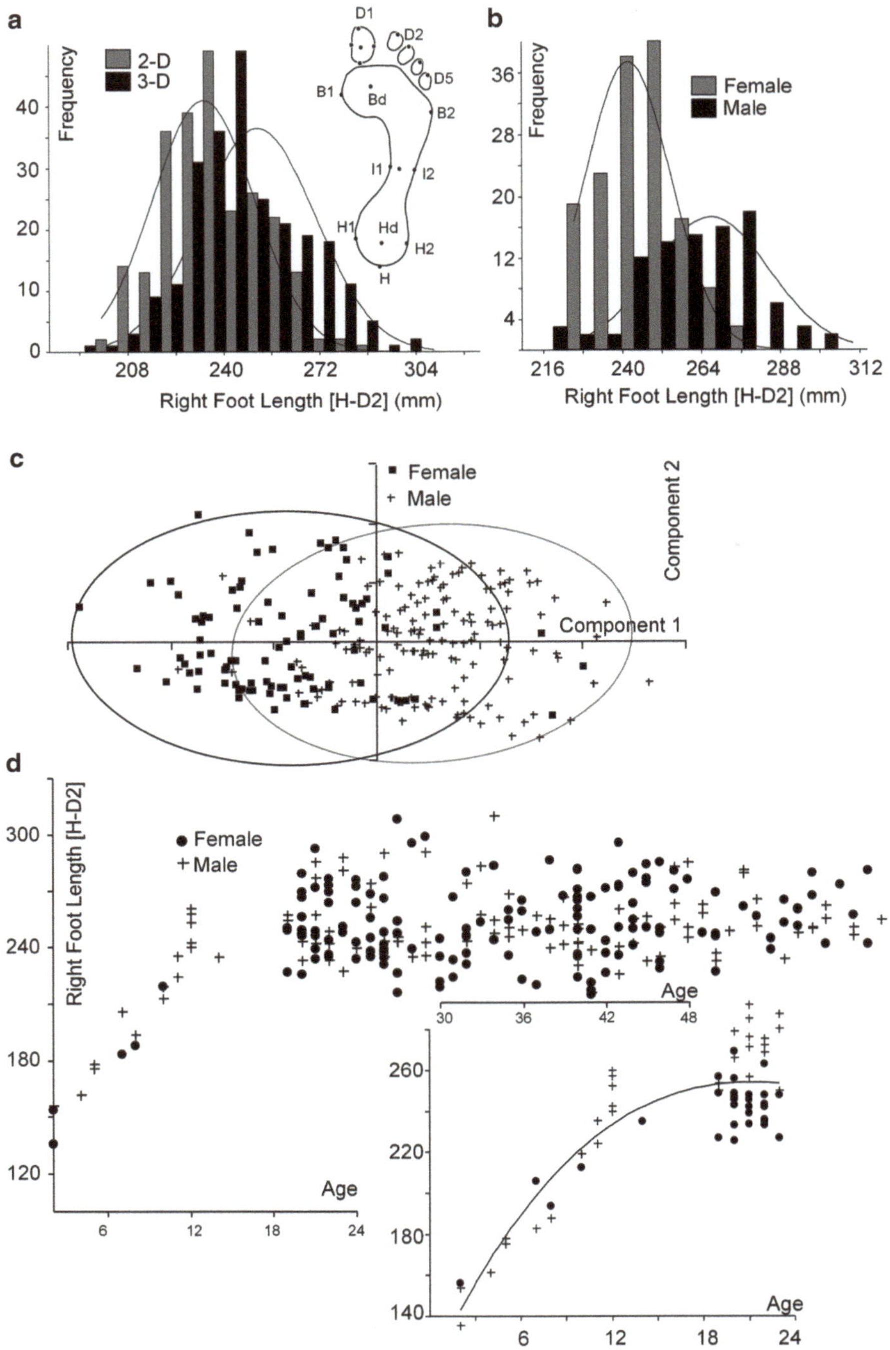

Fig. 6.2 Various data plots for the Bournemouth track data see Sect. 1.4 for details on data collection. (**a**) Frequency histogram of right foot length (heel to second toe) for two-dimensional and three-dimensional tracks. (**b**) Frequency histogram of right foot length (heel to second toe) comparing male and females. (**c**) First component of the Principle Components Analysis (PCA) showing separation of male and female subjects. (**d**) Variation in right foot length (heel to second toe) associated with age

Table 6.2 Selected data for the tracks within a study of modern habitually shod humans working at Bournemouth University in 2007

Variable		Female [150]		Male [93]	
Three-Dimensional Tracks		*Right*	*Left*	*Right*	*Left*
Length [H-D2]	X	241.66	242.2	266.7	267.8
	R	193.5–272.4	189.7–287.2	218.9–309.5	218.7–309.4
	σ	1.0661	14.6	17.7	16.2
Ball Breadth [B1-B2]	X	100.77	101.3	108.9	109.1
	R	75.3–117.9	76.9–116.7	81.1–133.8	81.5–131.5
	σ	0.529	6.5	8.9	8.4
Heel Breadth [H1-H2]	X	64.25	64.8	70.56	71.2
	R	47–90.8	49.0–81.8	49.8–95.5	50–99.9
	σ	0.49	5.8	6.9	6.6
Two-Dimensional Tracks		*Right*		*Right*	
Length [H-D2]	X	225.02		246.4	
	R	176–253		193–285	
	σ	12.44		15.3	
Ball Breadth [B1-B2]	X	87.8		96.6	
	R	12–101		79–113	
	σ	0.86		0.68	
Heel Breadth [H1-H2]	X	55		60.8	
	R	38–102		41–108	
	σ	1.43		1.73	

All measurements are in mm and quoted to two decimal places [X = mean; R = range; σ = standard deviation]. See Fig. 6.2a for landmarks and distance definitions

of 18.05 mm giving a statistically significant different ($t = 10.895$; $p < 0.0001$; Fig. 6.2a) Equally foot breadth is on average 12.5 mm less than for the three-dimensional track ($t = 17.221$; $p < 0.0001$). One would expect the three-dimensional tracks to be slightly larger due to widening of the track walls by the sides of the foot and due to the differences between walking and standing. Track 1 in the Appendix illustrates the issue very well, with a clear difference between the overall track and the true track (Table 1.1). These differences may be perhaps mitigated but are clearly not off-set by the greater compression of soft tissues around the standing foot resting on the hard non-compliant surface on which the two-dimensional track records were made. These results suggest in broad terms that a two-dimensional static track is perhaps the order of 7 % smaller on average in terms of length and perhaps as much as 12 % in terms measures of foot width.

Contrary to some previous work (e.g., Agnihotri et al. 2007; Krishan 2008a, b; Sen and Ghosh 2008; Fawzy and Kamal 2010; Nataraja Moorthy et al. 2014) there is no evidence in the population sampled using either the three-dimensional or two-dimensional track data for any bilateral asymmetry. Left tracks are marginally larger by the order of a millimetre (Table 6.2) but there are no statistically significant differences. Sexual dimorphism is evident within the sample (Table 6.2) with female tracks being typically between 7 and 9 % smaller than those made by male subjects although not in all cases (Right H-D2 $t=11.765$ $p<0.0001$; Right B1-B2 $t=7.035$ $p<0.0001$; Fig. 6.2b). A Principle Components Analysis (PCA) of 14 inter-landmark distances gives shows some sexual separation with over 50 % of the variance being explained by the first component (Fig. 6.2c), but it is also clear that there is a lot of overlap within the data. Similarly a two-class discriminant analysis based on 14 inter-landmark distances gives only 84.36 % correct sexual classification. There are no obvious differences in the shape or pattern of depth (i.e. plantar pressure) distribution between the sexes using the full 19 were landmarks between the sexes; in a PCA based shape analysis over 33 % of the variance is accounted for in both populations simply by the placement of the deepest point in the ball. Despite the presence of sexual dimorphism we have chosen to use pooled – male + female – data in building regression models between biometric variables and foot dimensions, because the sex of a track-maker is not normally known in a fossil case. Subjects with an age below 19 were excluded from these analyses since it is generally assumed that adult stature is normally achieved by 19 to 23 years old (Roche 1986). However the variation in right foot length with age in the whole sample is shown in Fig. 6.2d and a third order polynomial curve can be successfully fitted with a high degree of significance (Chi = 16,075, Akaike IC = 5.775, $R^2=0.77$, $F=6.78$, $p<0.0001$) to the 63 subjects 23 years old or younger. There is a slight decline in stature with age (Fig. 6.2d), but given that the age of a fossil track-maker is not normally known age was not factored into the adult regression models.

Typical regression models all significant at $p<0.0001$, are shown in Table 6.3 and a couple of key distributions are illustrated in (Fig. 6.3). Good correlations are observed between foot length and stature, with best regression coefficients being achieved using the heel to second toe measurement. Slightly better regression values and reduced errors are found in using the two-dimensional tracks and these give stature estimates that are between 36 and 68 mm larger than those based on three-dimensional tracks. The implication here is that regression models developed from two-dimensional tracks may over estimate stature when applied to data taken from three-dimensional tracks in the fossil record. Within the sample there is some variation in the length of the first toe however it does not correlate in a statistically significant fashion with foot length or stature. The variation appears to be more associated with subtle variations in the locomotor styles of individuals which do not appear to vary with age, body mass or sex in any systematic way. Some individuals appear to curl their toes slightly making more prominent circular pad impressions often linked to the proximal movement of sediment behind the toes, while others seem to place the toes in a more plantigrade fashion creating more elongated toe pads with less movement of sediment to the rear.

Statistically significant relationships were found between body mass and various foot dimensions the strongest of which was between body mass and foot breadth; although the strength of all those relationships is modest with high error values (Table 6.3). Combining height and body mass in the Body Mass Index (BMI) BMI does not appear to provide any clear value within these regression models, although a statistically significant but weak relationship was found between BMI and foot breadth. The

Table 6.3 Regression models derived from the Bournemouth track samples (see Sect. 1.4)

	Equation	r	R^2	Error b	Error a
Stature-3D					
RFL2	H=0.00577X+0.236	0.839	0.70	0.5072	0.0002
RFL1	H=0.00581X+0.186	0.799	0.63	0.0578	0.0002
RFL5	H=0.00676X+0.266	0.836	0.69	0.50	0.0002
RFHB	H=0.01580X+0.632	0.553	0.30	0.0565	0.0008
RFBB	H=0.01334X+0.302	0.643	0.41	0.068	0.0006
LFL2	H=0.00565X+0.259	0.831		0.051	0.0002
Stature-2D					
RFL2	H=0.00651X+0.1676	0.856		0.050	0.0002
RFL1	H=0.00673X+0.1049	0.851		0.053	0.0002
RFL5	H=0.82843X+5.5087	0.854		6.346	0.0323
RFHB	H=0.02257X+0.5448	0.556		0.061	0.0001
RFBB	H=0.01525X+0.29156	0.679		0.065	0.0007
Weight-3D					
RFL2	W=0.9793X−172.22	0.471		13.96	0.0556
RFHB	W=2.6815X−104.93	0.543		9.683	0.1450
RFBB	W=2.265X−160.93	0.567		12.43	0.1201
LFL2	W=0.960X−168.19	0.469		13.75	0.0546
BMI-3D					
RFBB	BMI=0.723X−49.156	0.297		4.608	0.0444

Most of the models shown are for the right foot. Only those valid at <0.0001 are shown. *RFL* right foot length, *RFHB* right foot heel breadth, *RFBB* right ball breath. Numbers refer to toe digits such that RFL2 is the distance from the heel to the tip of the second digit. See Fig. 6.2a for dimensions

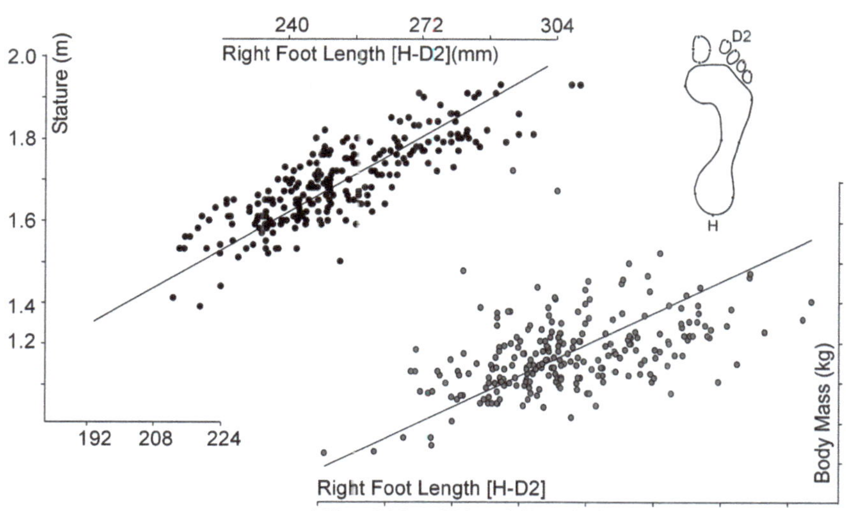

Fig. 6.3 Biometric relationships between right foot length (heel to second toe) for the Bournemouth data (see Sect. 1.4) and stature and body mass. Regression equations are given in Table 6.3

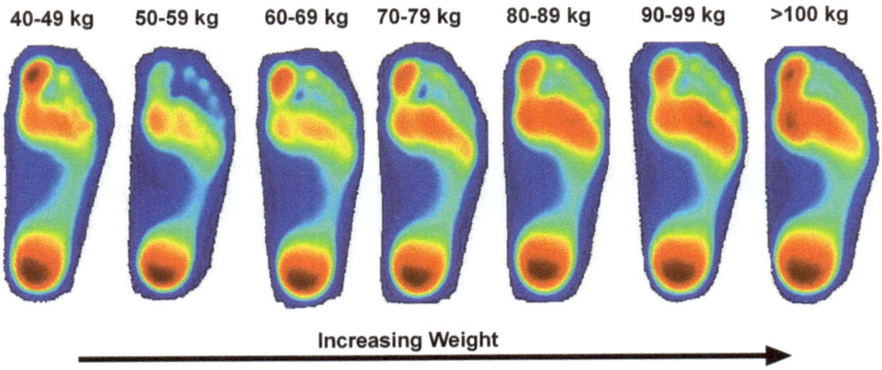

Fig. 6.4 Mean tracks for different weight classes based on pooled male and female data from the Bournemouth track data (see Sect. 1.4). Each subject left four tracks – rights and lefts – these have been combined into a mean for that subject and then used to create the weight means. All means have been created in Pedobarographic Statistical Parametric Mapping (pSPM) software described in Sect. 2.7

presence of some type of relationship is perhaps unsurprising since variation in BMI is known to correlate with plantar pressures and gait parameters (e.g., McGraw et al. 2000; Lai et al. 2008; Mignardot et al. 2010; Ko et al. 2012). No correlation was found between the Clarke Ankle (Fig. 2.12) and weight which one might expect; larger subjects potentially having flatter feet. There are no statistically significant relationships between track depth and either body mass or stature (see Sect. 5.3), although deeper tracks are generally associated with heavier subjects. The distribution of depth within a track and its variation with weight and BMI were explored by creating a four track mean for each subject within Pedobarographic Statistical Parametric Mapping (pSPM; see Sect. 2.7) combining both right and left tracks. Figure 6.4 shows a series of mean tracks for a range of weight classes; as subject weight increases there is a tendency for enhanced forefoot depths, with more pronounced merged toe impressions and a wider area of maximum depth under the ball without visible metatarsal head impressions. This result is not surprising, since most men and women carry 'additional weight' at the front which may potentially move their centre of gravity forward. Further work is needed to explore the impact of BMI on track topology with important implications for trip prevention in more overweight individuals especially with increasing age.

Faced with the plethora of foot related studies, including those just presented here, it is genuinely difficult for the geoarchaeologist to proceed in selecting the approach or regression model with which to make biometric inferences from a series of tracks. We can use Harry's Trackway (see Sect. 1.4) to help illustrate this point. Table 6.4A shows the variability in track measurements along a single trail (Fig. 2.13) due to variations (as we saw in Sect. 5.3) in the substrate over which the track-maker walked. Taking a selection of the published multipliers and regression models, we can apply them to Harry's Trackway in order to illustrate the range of possible statures for 'him' (Table 6.4B). Stature estimates vary from as little as 1.35 m to over 1.73 m, with a mean of 1.64 m; the range of possible inferences is

Table 6.4 A. Track length measurements for Harry's Trackway at Walvis Bay (Namibia). B. Measures of stature and body mass based on Harry's Trackway

A			Min	Max	Mean	Median	Standard deviation
	Harry's Trackway – foot right length		207.10	274.35	234.44	233.1	12.51
	Harry's Trackway – foot left length		212.89	252.85	230.92	231.46	10.20
B		**Ethnicity/Race**	**Min**	**Max**	**Mean**	**Median**	**Standard deviation**
1	Toppinard (1877) 16 %	Generic	1.38	1.80	1.52	1.52	0.07
2	Toppinard (1877) 14 %	Generic	1.57	2.05	1.73	1.74	0.08
3	Barker and Scheuer (1998) 18.90 %	Generic (UK-US)	1.22	1.59	1.34	1.35	0.06
3	Barker and Scheuer (1998) 14.10 %	Generic (UK-US)	1.48	1.93	1.63	1.63	0.08
4	Dingwall et al. (2013)	African (Daasanach)	1.56	1.81	1.65	1.64	0.04
5	Webb et al. (2006)	Aborigines	1.45	1.89	1.6	1.6	0.07
6	Tuttle et al. (1990)	American Indian	1.49	1.94	1.64	1.64	0.08
7	Krishan (2008b)	Indian (Gujjars)	1.65	1.88	1.72	1.72	0.05
8	Robbins (1986)	Generic	1.53	2.01	1.69	1.69	0.08
9	Agnihotri et al. (2007) male	Mauritius	1.58	1.85	1.66	1.66	0.05
10	Agnihotri et al. (2007) female	Mauritius	1.55	1.79	1.63	1.63	0.04
11	Fawzy and Kamal (2010) H-D1	Generic (Egyptian)	1.6	1.81	1.67	1.67	0.03
12	Sen and Ghosh (2008)	Indian (Rajbanshi and Meche)	1.51	1.84	1.62	1.62	0.06
13	Reel et al. (2012) H-D1	Generic (UK)	1.57	1.83	1.66	1.66	0.04
14	Uhrová et al. (2013) H-D1	Caucasian (Slovakian)	1.58	1.9	1.69	1.69	0.05
15	Ukoha et al. (2013) H-D1	African (Nigerian)	1.63	1.84	1.70	1.70	0.03
16	Kanchan et al. (2012) H-D1	Generic (Indian)	1.6	1.97	1.72	1.72	0.06
17	Nataraja Moorthy et al. (2014) H-D1	Indian (Tamils)	1.67	1.84	1.73	1.73	0.03
18	Atamturk and Duyar (2008)	Turkish	1.56	1.91	1.67	1.67	0.06
19	Adams and Herrmann (2009) males	Generic	1.58	1.82	1.66	1.66	0.04
20	Adams and Herrmann (2009) females	Generic	1.55	1.78	1.62	1.62	0.04
21	Nataraja Moorthy et al. (2013a, b) H-D1	West Malaysia	1.6	1.93	1.7	1.7	0.06
22	Pawar and Pawar (2012) males	Generic (Indian)	1.55	1.98	1.69	1.69	0.07

(continued)

Table 6.4 (continued)

B		Ethnicity/Race	Min	Max	Mean	Median	Standard deviation
23	Pawar and Pawar (2012) females	Generic (Indian)	1.58	1.79	1.65	1.65	0.04
24	Kanchan et al. (2008)	Indian (Gujjars)	1.55	1.74	1.61	1.61	0.03
25	Hairunnisa and Nataraja Moorthy (2014)	East Malaysia	1.52	1.85	1.64	1.64	0.06
26	Table 6.3 this volume	Generic (UK)	1.43	1.81	1.57	1.57	0.01
	Master means		1.34	1.73	1.64	1.64	

reinforced by reference to Fig. 6.5a. Interestingly the population specific regression models tend in general to under-estimate stature, while the more generic relationships give higher values. The estimates of body mass show an even greater level of variation – in some estimates Harry is positively undernourished in others he is clinically obese! This reinforces the challenge of making reliable body mass estimates from tracks. In one case the relationship of Fawzy and Kamal (2010) gives an estimate of body mass based on the right foot of just 63 kg compared to one of over 134 kg when the left foot is used! The point here is that while many of these regression models are clearly not applicable there is a lot of latitude with respect to which one to choose for an unknown track-maker and the results can be very varied.

The challenge for those studying ancient trackways becomes more manageable if there is clarity of aim and purpose in making the inference. If it is a forensic one, then selecting the most appropriate regression model for the likely, or most probably, population from which the track-maker comes is critical. Being aware of the potential here for sampling and methodological artefacts within these models is essential. Where evidence of age and sex can be gleaned from other sources such as witness statements, then they do add to the superiority of the regression models as Atamturk and Duyar (2008) argued. If the population is not known, for example in an ethnically diverse population, the estimates obtained are likely to be only broad. In archaeological contexts the aim is usually to simply provide some descriptive colour about the track-maker. In which case the accuracy of any inference about stature or body mass is not critical and the application of a generic multiplier like the 15 % of Toppinard (1877) is in truth probably as an accurate approach as any (e.g., Roberts et al. 1996; Avanzini et al. 2008). Webb et al. (2006) documenting the late Pleistocene prints of Willandra Lakes in southeast Australia provides a model approach where appropriate anthropological data is available to allow a more sophisticated solution. They used historic anthropometric data on native aborigines to develop a bespoke regression model, while the sample was modest (N = 126) it is likely to have been from a similar racial/ethnic pool as the track-makers and therefore has the potential to give good stature estimates.

These challenges are exacerbated when dealing with extinct hominins for which there are no robust (or complete) empirical models linking the relative size of different body parts. In addition identifying the likely track-maker is often the aim here and estimates of stature and body mass may be of greater significance than in other contexts (Dingwall et al. 2013; Ashton et al. 2014). Dingwall et al. (2013) provides a

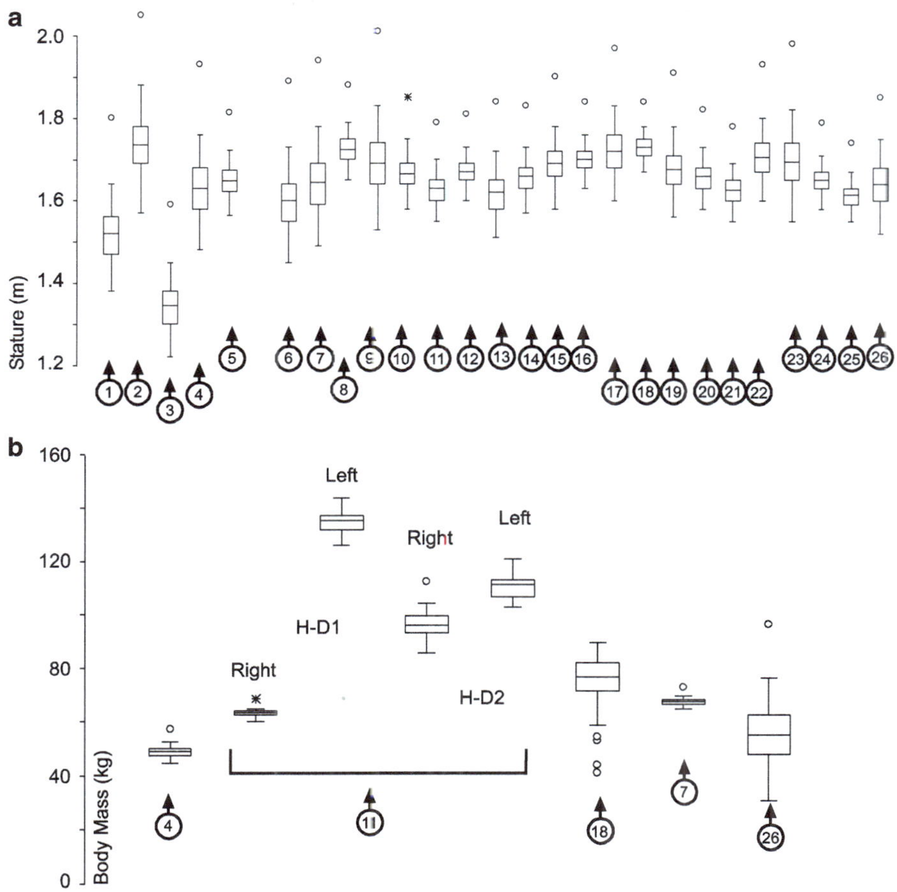

Fig. 6.5 Stature estimates based on Harry's Trackway (see Sect. 1.4). In all cases unless otherwise stated foot lengths are from the heel to first toe. The circled numbers correspond to the method indicated in Table 6.4. (**a**) *Box plots*, with statistical outliers also shown using a range of different regression models. (**b**) Body mass estimates

model approach for this type of scenario applied in this case to the 1.5 Ma Ileret footprints, which has been followed recently by Ashton et al. (2014) although the relevance of the reference population may be more questionable. Dingwall et al. (2013) used a controlled field experiment to collect actual footprint lengths, walking speeds and body mass/stature information from a small sample of habitually unshod Daasanach subjects who currently live at Ileret (Kenya). Regression models were used to interpret the fossil tracks on the basis that the environmental and body mass of current inhabitants would have been similar to the ancient track-makers. While this is assumptive across species there are few alternatives. On this basis Dingwall et al. (2013) were able to confirm the conclusion of Bennett et al. (2009) that the likely track-maker was *Homo erectus*, although the data does not rule out the possibility of the tracks being made by a male *Paranthropus boisei*. However we need a word of

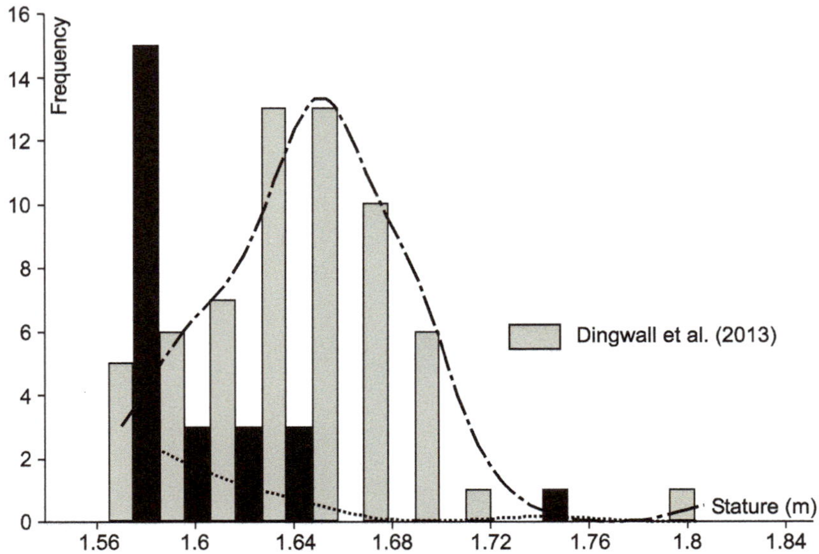

Fig. 6.6 Two different stature estimates for Harry's Trackway (see Sect. 1.4) using the regression equations of Dingwall et al. (2013; N = 38) and one derived here (N = 31) based on different samples of the Daasanach at Ileret in northern Kenya

caution here. The lead author also sampled the Daasanach population in 2009 and while the sample was slightly smaller (N = 31 compared to N = 38) the regression model obtained is different [$H = 0.00489FL + 0.39867$] as illustrated in Fig. 6.6 when applied to Harry's Trackway. While the sampling of Dingwall et al. (2013) was no doubt more robust than that of the author's it is a cautionary tale in building empirical models from small samples. Identifying the most appropriate modern analogue population for and extinct hominin species is challenging and those native to a site now may not be the most appropriate analogue. Tuttle et al. (1990) used the length to width ratio of the Laetoli tracks to find a modern analogue with a similar ratio finally selecting the Machiguenga Indians of Peru. In a study of 69 subjects they not only analysed the gait but also the foot length to stature relationships developing a regression model which gave a mean stature estimate of 1.22 m for the –1 and 1.41 m G-3 trails. The work provides a bench mark in how to select an appropriate reference population and contrasts with that of Dingwall et al. (2013) where the assumption is that today's indigenous population is representative of that in the past.

6.3 Inferring Age

Inferring the presence of children within an assemblage of human tracks is relatively common (e.g., Roberts et al. 1996; Lockley et al. 2008; Schmincke et al. 2010; Ashton et al. 2014) and usually based on the presence of small tracks. According to

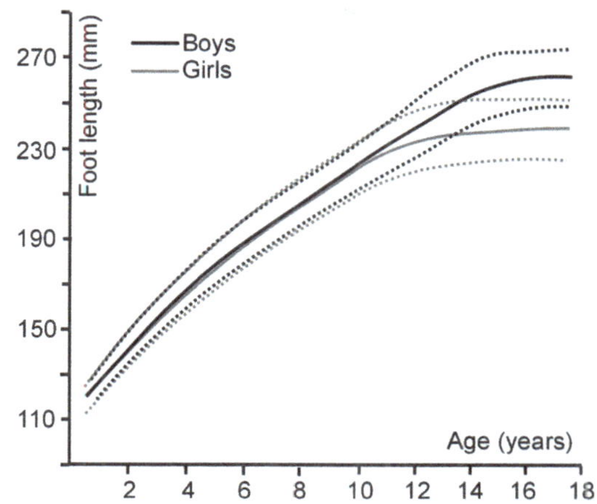

Fig. 6.7 Generic growth curve showing variation in foot length with age (Modified from Ashton et al. (2014))

Roche (1986) adult body proportions are normally established by 19 years, although 10 % males may continue to grow into their early twenties (Roche and Davila 1972). It is widely assumed that body proportions are then fixed but some authorities suggest that they may vary throughout life (Perissinotto et al. 2002; Dangour 2003; Kanchan et al. 2010b) with a decline in stature likely to be apparent after the fifth decade (Friedlaender et al. 1977). Typical growth curves for children's feet are shown in Fig. 6.7 but are very dependent on the individual's sex as well as their race/ethnicity and crucially the levels of nutrition (e.g., Anderson et al. 1956; Hill 1958; Malina et al. 1973; Stavlas et al. 2005; Grivas et al. 2008; Bosch et al. 2010; Muller et al. 2012). The dimensions of adolescent feet overlap with those of small adults especially those of young women where there is marked sexual dimorphism within a population. Consider the growth curve for the Bournemouth data (see Sect. 1.4; Fig. 6.2d); there is a clear overlap in the size of subject's tracks after the age of 10. Therefore inferring the presence of children's tracks should be a matter of caution. The challenge is illustrated by the children's tracks described from Walvis Bay (Namibia; Bennett et al. 2014). The smallest group of tracks have a mean length of just 114 mm. In a typical study of 7788 children by Muller et al. (2012) suggested that a 1 year old had a mean foot length of 13.07 ± 1.59 cm rising to 24.4 ± 2.96 at the age of 13. The sizes recorded in Namibia are much smaller, yet clearly human tracks. If one uses published growth curves (Davenport 1932; Anderson et al. 1956) and compensates for an assumed lower level of nutrition the tracks may have been made by individuals as young as 3 or 4 years old (Bennett et al. 2014). This is of course conceivable given the association with larger tracks made presumably by women or adolescents, but it is still a challenging observation. The problem, as at most archaeological sites, is that there are no population specific growth curves with which to make a valid comparison. It is generally assumed that children's feet have an adult form, if not size, around the age of 6 years. A number of studies suggest that a child's foot elongates as it grows and that the arch becomes fully established in the first 6 years of growth although this may vary with the child's sex

(Bertsch et al. 2004; Stavlas et al. 2005; Grivas et al. 2008; Bosch et al. 2010; Muller et al. 2012). Given a sufficient number of tracks made by young children it may be possible to pick out these changes in three-dimensional tracks, but it has yet to be attempted.

6.4 Fossilised Locomotion? Inference of Speed and Gait

Estimating the speed at which a track-maker was travelling is another logical inference to make from a series of tracks. Theoretically for a given set of body proportions an increase in speed should normally result in an increase in the subject's stride length, effectively an increase in the ground being covered in a unit of time. Equally two individuals with different stature and limb lengths running at a similar pace should yield trackways with different stride lengths. In this scenario the individual with the shorter limbs will need to take a higher frequency of steps in a given unit of distance to keep up. The point here is that the correlation between stride length and velocity is dependent on the biomechanical geometry of the track-maker as well as the style or mode of gait. An individual moving with for example with a bent-hip and bent-knee posture will cover the ground in a different way than one walking with an upright posture. The measurement of stride length and step length is covered Sect. 2.5 (Fig. 2.11). According to Alexander (1984) a hominin walking with speed (v) and taking strides (λ) will be related via some form of mathematical function F, such that:

$$\lambda = \mathrm{F}(v) \tag{6.1}$$

In theory the function F will be species specific, varying between hominins of different sizes. Charteris et al. (1982) in making inferences on speed from the Laetoli tracks argued that the issue of body size can be accommodated for by introducing stature (h) on both sides of the equation to give relative step length and relative speed, such that:

$$\lambda / h = \mathrm{F}(v / h) \tag{6.2}$$

According to Alexander (1984) there is little theoretical foundation for this, however regression models can be derived for a given population as illustrated by Charteris et al. (1981, 1982). These empirical relationships should hold for that population, in much the same way as we have seen that relationships between stature and foot length do. The problem comes when one applies such models to extinct hominins where stature is unknown. Alexander (1984) explored this issue by introducing data with greater variability in stature, namely containing children (short humans) and adults. He argued during earlier work (Alexander 1976) that the problem of inferring speed could be viewed as one of dynamic similarity. Two pendulums of different lengths can move in a dynamically similar fashion. Our two systems will have an equal Froude Number v^2/gl, where v is the speed of the system, g is the gravitational constant and l is the length characteristic of the system. In the case of our pendulum the length of the wire suspending the weight would be the most appropriate measure, in humans it might be hip height. We can re-write Eq. 6.2 such that:

$$\lambda / h \approx \mathrm{F}\left(v / \sqrt{gh}\right) \tag{6.3}$$

While this may be a more theoretically correct formulation (Alexander 1984) it still requires an empirical model on which to determine the value of F, just that in theory this model should be applicable across a much greater range of body lengths, something which Alexander (1984) confirms from his study of children and adults, although there is greater divergence at high speeds. In his original work Alexander (1976) derived on the basis of beach-based experimental data a power law with which to estimate speed:

$$\lambda / h \approx 2.3\left[v^2 / (gh)\right]^{0.3} \tag{6.4}$$

$$v = 0.25 g^{0.5} \lambda^{1.67} h^{-1.17} \tag{6.5}$$

This approach has been widely used for both humans and bipedal dinosaurs (Sellers and Manning 2007). Ruiz and Torices (2013) tested this using data from elite athletes and from their own beach studies to confirm and refine the basic power law described by Alexander (1976) across a wide range of individuals.

$$v = 0.794\ \lambda^{1.67} \tag{6.6}$$

They do note however that speed is non-unique for a given stride length, but also a function of the length of the sporting event. This highlights the importance of stride frequency in these determinations which cannot be determined from fossil trackways alone and the need for caution is clear.

Calculating speed from fossil tracks has attracted publicity in the case of the Willandra Lakes track site (Webb et al. 2006), where several trackways were made by individuals running (see Sect. 3.2.7). The original estimates based on an approach similar to that of Charteris et al. (1981) gave in one case speed of 10.3 ms^{-1} which would give an Olympic sprinter a competitive race (Webb 2007; McAllister 2011). Ruiz and Torices (2013) estimate using their power law that more realistically this was in the order of 7.15 ms^{-1} which is still fast, but not unrealistically so. It supports their general conclusion that the approach of Charteris et al. (1981) may over estimate a track-maker's speed.

Dingwall et al. (2013) provides a recent illustration of this approach in their interpretation of the Ileret tracks. Using a sample of 38 adult Daasanach subjects they computed regression equations using both the approach of Charteris et al. (1981; speed versus stride length-to-foot length ratio) and that of Alexander (1984; Froude Number versus dimensionless stride length). In the latter case they used the height of the greater trochanter (top of the thigh) as the measure of limb length. Both approaches gave good correlations suggesting that the track-makers who made the main trackway on the upper surface at Ileret was moving slowly (0.45–1.0 ms^{-1}) and consistent with the slippery and uncertain substrate, although one trail with just two prints on the lower surface gave an estimate of 2.2–2.7 ms^{-1} which is more consistent with a fast walk or a slow run. These estimates proved robust across a range of stature and hip height estimates.

The application of such approaches to the Laetoli footprints has generated by contrast a much greater amount of literature and is intimately tied up in the debate over the degree to which these tracks show modern characteristics. Comparing the stride lengths with modern humans gives speeds which vary from 0.56 to 0.75 ms^{-1} depending on the Trail (G-1 versus G-2) and the precise method and modern regressions models used (Charteris et al. 1981, 1982; Alexander 1984; Tuttle 1987; Reynolds 1987; Tuttle et al. 1990). These walking speeds are consistent with very slow walking rates, well-below normal human preferences, being described as a 'gentle stroll' by some workers (Charteris et al. 1981, 1982). Sellers et al. (2005) obtained a similar range (if slightly elevated) of velocities by applying a biomechanical simulation to the trackways. Raichlen et al. (2008) demonstrated that higher rates of speed could be obtained if one used regression models based on chimpanzees walking with a bent-hip-bent-knee gait rather than an up-right posture adding to the debate about the locomotor styles of the Laetoli track-maker. This debate does illustrate the dependence of the result on the choice of an appropriate regression model.

The debate around the Laetoli tracks highlights the final issue here which is fundamental to the interpretation of ancient hominin tracks; namely the degree to which the locomotor styles differ and ultimately the degree to which locomotion may have been a factor in our own evolution. Put another way did the track-maker at Laetoli have a modern foot and style of locomotion? If not when did this adaptation take place before or after the transition from the *Australopithecus* to *Homo* associated with many other changes in human anatomy (Bramble and Lieberman 2004). The debate over the Laetoli tracks has been long, often bitter and frequently based upon differing interpretations of single tracks showing features that support a particular writer's hypothesis. Essentially as stated by Meldrum et al. (2011) the debate has been broadly polarised between two alternate hypotheses which can be generalized as follows:

1. The tracks are fundamentally indistinguishable from those of modern humans with an essentially modern foot anatomy and up-right gait leading to a medial longitudinal arch, lateral-to-medial force transfer and push-off by the hallux (e.g., Day and Wickens 1980; Charteris et al. 1981; Alexander 1984; Lovejoy 1988; Tuttle 1987; Tuttle et al. 1990, 1991; Musiba et al. 1997; Schmid 2004; Harcourt-Smith 2005; Sellers et al. 2005; Kimbel and Delezene 2009; Raichlen et al. 2011; Crompton et al. 2012).
2. The tracks indicate a track-maker with a foot architecture that is manifestly distinct from modern humans showing some but not all features of modern foot and gait, with some workers arguing for a bent-hip, bent-knee style of locomotion and a mosaic (or intermediate) range of foot characteristics (e.g., Stern and Susman 1983; Susman et al. 1984; Deloison 1991, 1992; Clarke 1999; Meldrum 2000, 2002, 2004a, b; Berge et al. 2006; Meldrum and Chapman 2007; Meldrum et al. 2011).

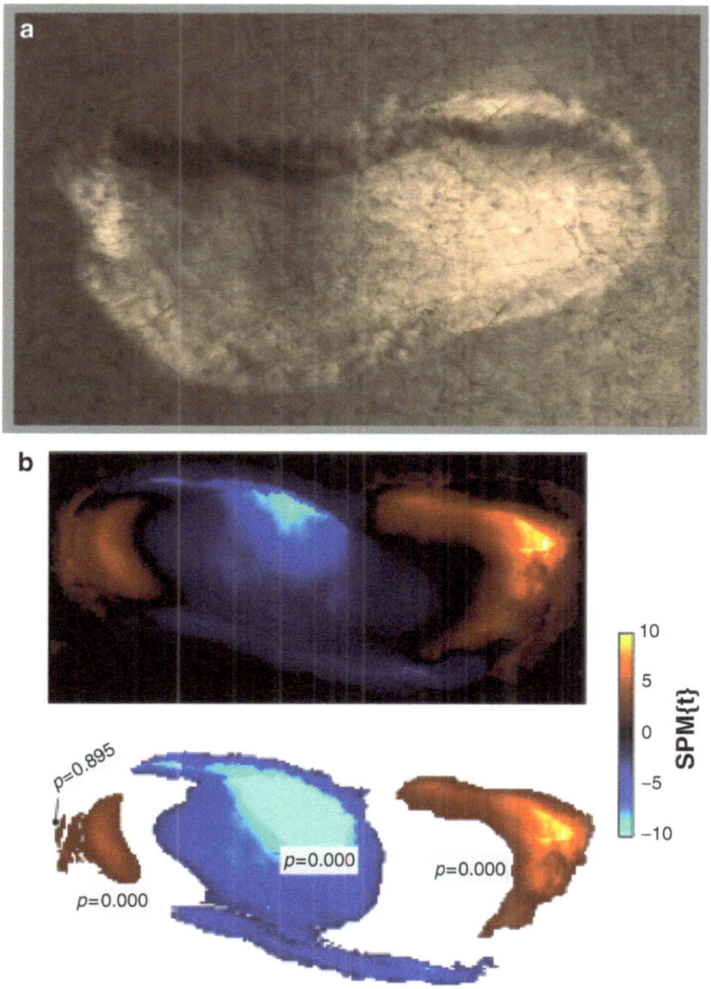

Fig. 6.8 Laetoli tracks. (**a**) Mean track created from 11 prints within the G-1 Trackway using Pedobarographic Statistical Parametric Mapping (pSPM) software described in Sect. 2.7. (**b**) Pairwise Statistical Parametric Map (SPM) showing those areas of significant difference between the Laetoli mean and a mean made of a hundred modern human tracks. Areas of statistically significant difference are indicated (Modified from Crompton et al. (2012))

Perhaps the most definitive contribution to this debate in recent years is that provided by Crompton et al. (2012) who applied a whole-foot analysis to derive measures of central tendency from the G-1 Trackway and compared this to modern human tracks and pressure records. A mean of the tracks in the G-1 trackway (Fig. 6.8a) shows: (1) the heel impression is substantially deeper than the forefoot; (2) there is a continuous depression under the region of the metatarsal heads across the whole width of the foot; (3) there is a raised area under the medial longitudinal arch which does not extend into the lateral mid-foot; and (4) there is a clear impression under the toes. When compared statistically to the modern human foot (Fig. 6.8b) you can see that the Laetoli mean track has a medial longitudinal arch but that it is shallower than the modern print; the toes of the modern mean are general deeper especially around the first toe; the impression of the heel in the modern print is also more pronounced; and the lateral margin of the foot is shallower in the modern print. Interestingly the mean of a series of unshod tracks made by a sample Daasanach from northern Kenya mitigate some of these differences such that the toes are more fanned out, and the medial longitudinal arch is more subdued and the mid-foot region broader. This was used by Crompton et al. (2012) to argue that the Laetoli track-maker was essentially similar in foot anatomy and functional gait to modern humans and that mid-tarsal breaks which were considered to be a primitive trait by some workers (e.g., Meldrum et al. 2011) were not present. In fact Bates et al. (2013b) has shown that mid-foot mobility is a feature of some modern humans and that there is greater statistical overlap in terms of mid-foot mobility between modern humans and great apes than often implied. The key point in this study is that it confirms that gait was essentially similar to modern humans and that differences across the *Australopithecus* to *Homo* transition are not pronounced, a theme which we will return to in the next section.

Plantar pressure is known to vary with walking speed (Rosenbaum et al. 1994; Segal et al. 2004; Taylor et al. 2004; Pataky et al. 2008) and consequently one might expect this to be manifest within the topology of a track. According to Brown (1999) one can deduce not only the speed at which an individual is moving but also subtle variations in their motion by the deformation structures produced both around the margins and within the base of a track. Figure 6.9 shows the results of a simple experiment in which a single male adult (1.69 m, 69 kg) made a line of four tracks in a long jump pit walking at three different speeds (1.25, 1.68 and 1.74 ms^{-1}). The frequency of deformation structures associated with brittle failure and the occurrence of thin sediment slabs moving between the foot and track base visibly increases with speed. A similar experiment conducted at a local beach with a single female subject (1.45 m, 52 kg), this time using average tracks computed using Pedobarographic Statistical Parametric Mapping (pSPM; see Sect. 2.7) shows how the track typology of this individual changes with increasing walking speed (Fig. 6.10). In this case the heel becomes more rounded, the mid-foot becomes narrower and the contact area under the ball of the foot reduced and the toes become more deeply impressed. The stronger longitudinal arch with higher walking speeds is consistent with some plantar pressure observations (e.g., Segal et al. 2004; Taylor et al. 2004; Pataky et al. 2008) however it is impossible to generalise from these subject specific examples and further research is needed in this area. The movement of thin slabs of sediment within the base of the track within the shear zone formed between the foot and the track base is a feature of tracks made in more sandy substrates, especially where the surface layer has dried slightly with respect to the sub-layer. The frequency with which this occurs does in some cases appear to increase with walking speed and is a noticeable feature of the speed experiments described above. These slivers of sand along with other topological characteristics

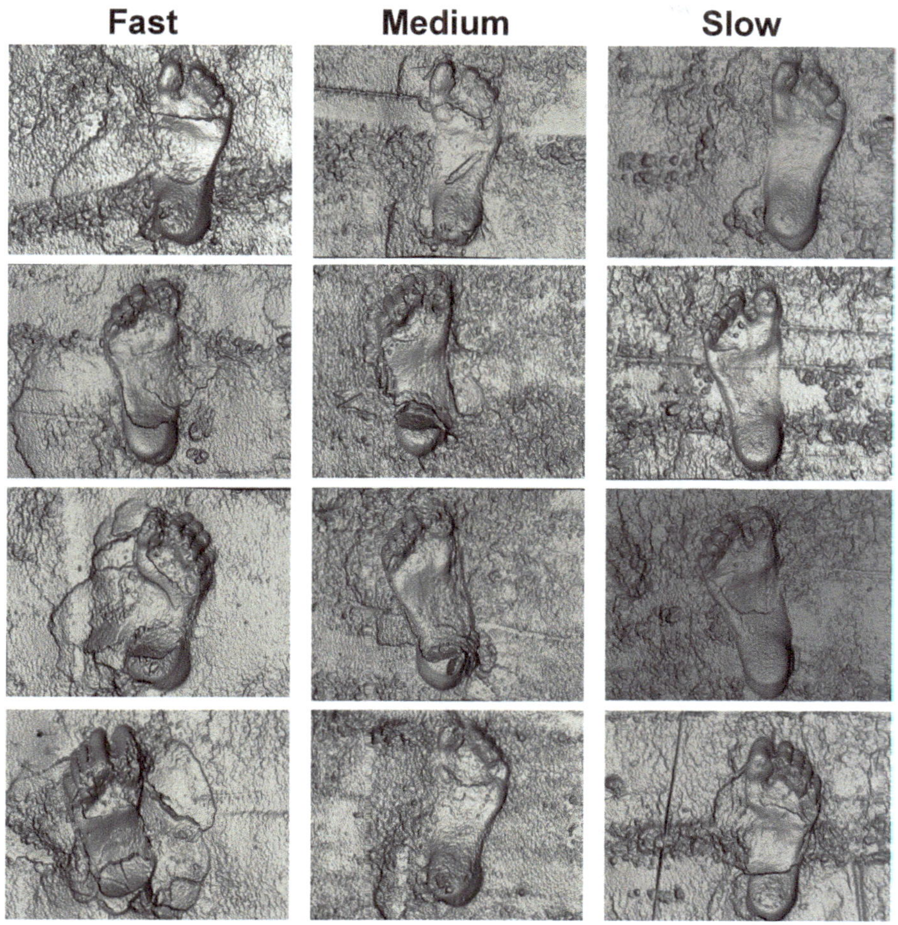

Fig. 6.9 A series of optical laser scans of three trackways each of four tracks created by a single individual moving at a different walking speed in each case. Note the increased deformation in the tracks under fast walking

can be studied as micro-scale tectonics revealing information about the patterns of applied shear stress. For example, Fig. 6.11a–c show variations on this theme for one subject while in Fig. 6.11d there are individual slivers associated with specific metatarsal heads allowing the relative sequence of motion to be determined; a point which is also true of Track 7 in the Appendix. In other cases there are multiple slivers (Fig. 6.11e) representing successive phases of failure and foot motion (Brown 1999) where these cross-cut the sequence of deformation is clear. Take the case in Fig. 6.11f; here a sliver associated with the lateral metatarsal heads is subsequently cross-cut by one beneath the second or first metatarsal head. These slivers are very thin and contrast with the larger structures generated when the foot 'digs' into the track base causing it to fail more generally (Fig. 6.12a, b). Importantly these features

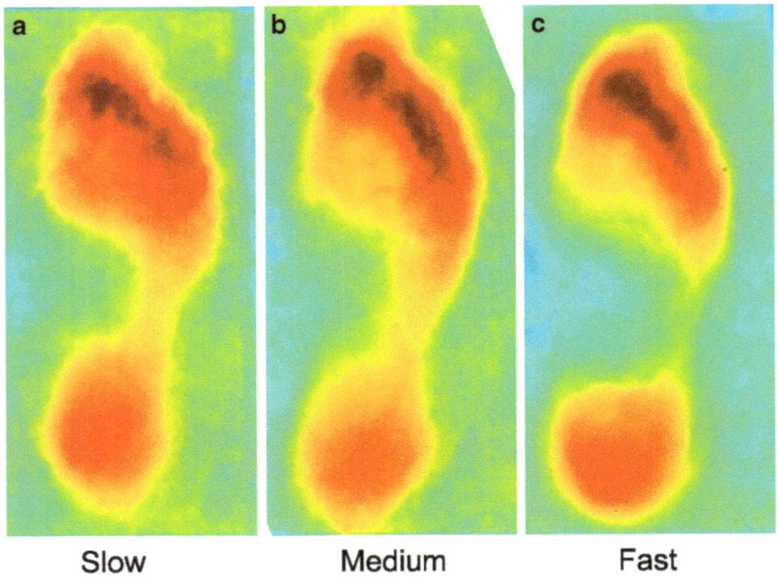

Fig. 6.10 Three mean tracks created using Pedobarographic Statistical Parametric Mapping (pSPM) software described in Sect. 2.7 for three trackways of 20 tracks created by a single individual walking at three different speeds on a sandy beach

originate from the ball of the foot but can terminate in the mid-foot (Fig. 6.12c; Track 8, Appendix). These are not necessarily indicative of mid-tarsal breaks as suggested in some work (Meldrum et al. 2011) but more a feature of sediment mobility in the micro-shear zone formed between the foot and a more consolidated track base. Another feature of these tracks is the orientation of the toes (Fig. 6.12d–f) which are indicative of changes of direction and the toes are typically more extended at high walking speeds. The varying toe orientations result in horizontal force vectors being directed at the track walls causing widespread deformation.

6.5 Evolution and Foot Function

In the previous section we touched on the role of hominin tracks as a source of evidence in our understanding of human evolution, describing the work of Crompton et al. (2012) who found little difference between the tracks at Lateoli and those of modern humans. In this section we amplify this theme by drawing on data and observations made throughout this book to do so. The appearance of Early African Homo (i.e. *Homo erectus/ergaster*) is marked by dramatic changes in body proportions from those typical of the best-known skeletons of *Australopithecus* (Ruff and Walker 1993; Anton 2003; Bramble and Lieberman 2004). The

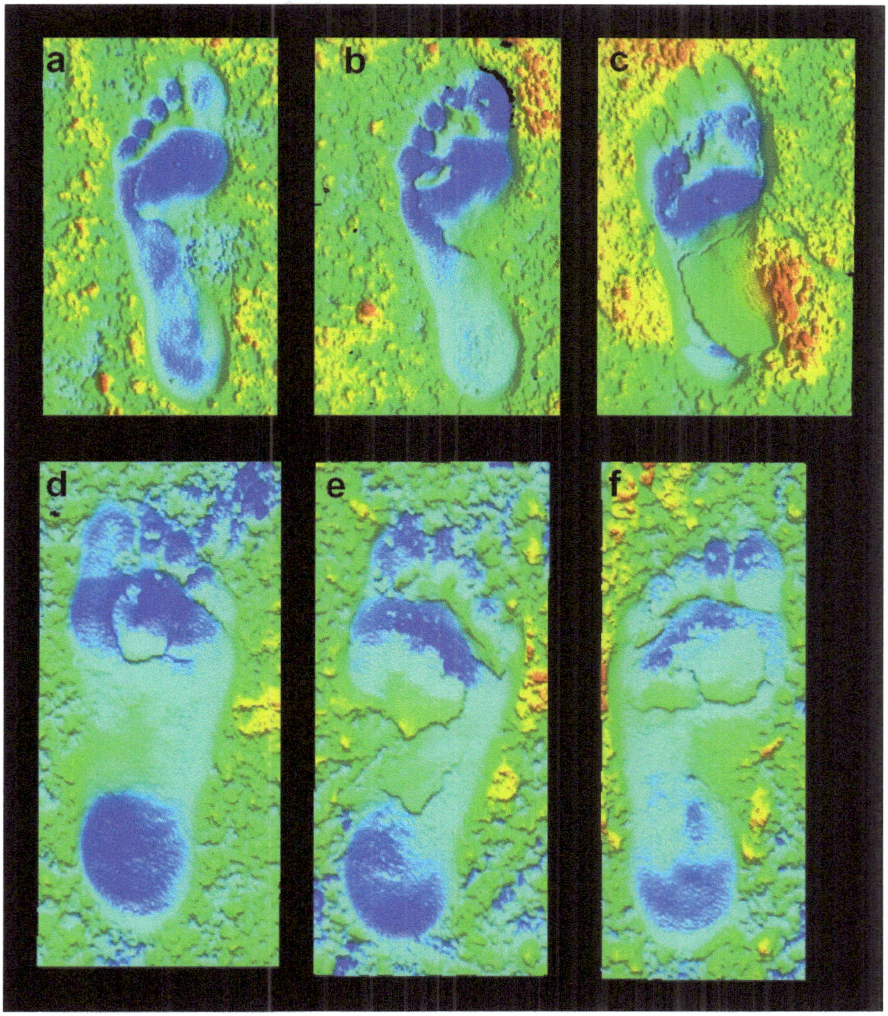

Fig. 6.11 A selection of scanned tracks showing various tectonic deformation features. See text for detailed discussion

long-trunked short-legged build of *Australopithecus* is replaced by a relatively short trunk and long legs, which is generally linked to the appearance of long distance striding bipedalism (Bramble and Lieberman 2004). It has been assumed that this was accompanied by the appearance of a human-like propulsive mechanism in the foot, with toe-off forces exerted primarily by the hallux (Bramble and Lieberman 2004). Fossil tracks, while comparatively rare in the geological record, offer a potential source of evidence with which to explore this idea. The alternative and more traditional solution is to use fossil foot bones but these are few and far between and

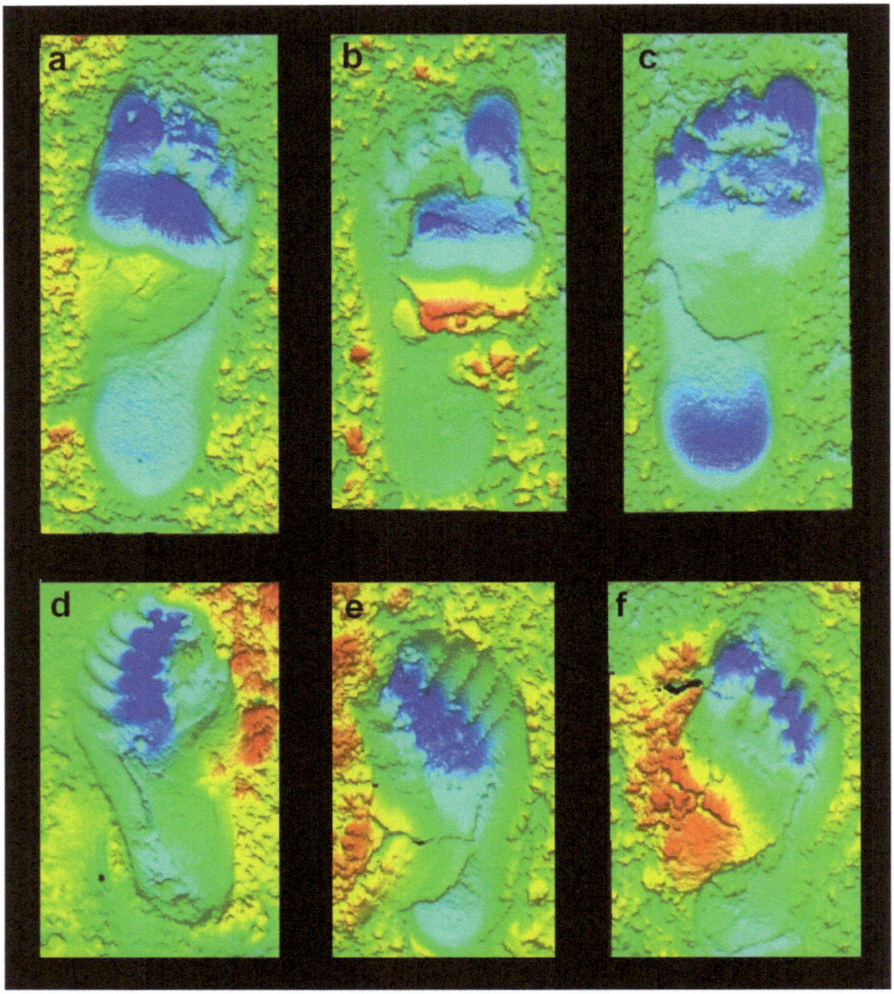

Fig. 6.12 A selection of scanned tracks showing various tectonic deformation features. See text for detailed discussion

only partial fossil feet are known prior to 2 Ma leading to a diverse, often contradictory, set of interpretations of hominin foot-function prior to the appearance of the genus *Homo* (e.g., Stern and Susman 1983; White and Suwa 1987; Deloison 1991). The discovery of the Laetoli tracks (3.66 Ma, Tanzania) in the late 1970s (Leakey and Hay 1979) attributed to *Australopithecus afarensis* (White and Suwa 1987), focused research on contrasting the locomotor mechanics of *Australopithecus afarensis* with those of modern *Homo sapiens* and as we saw in the previous section interpretations vary (White and Suwa 1987; Stern and Susman 1983; Crompton et al. 2012). The discovery of the Ileret (Kenya) tracks in 2009 and attributed to

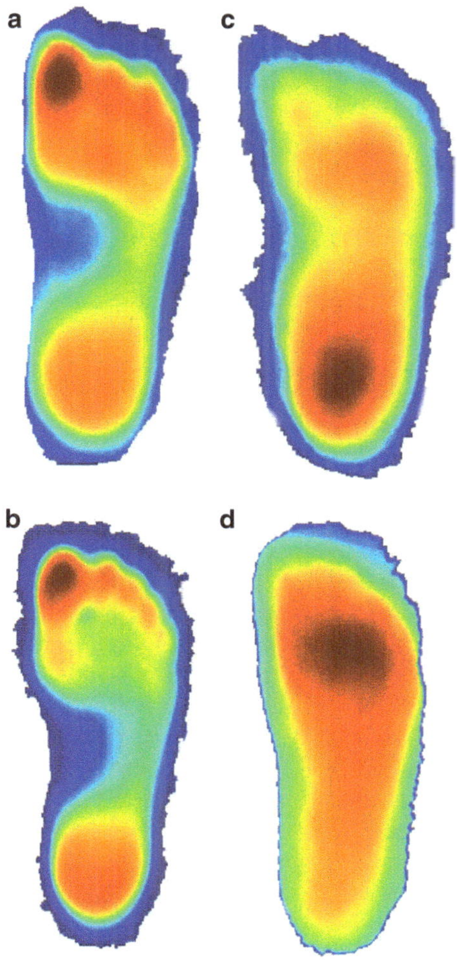

Fig. 6.13 Comparison of mean tracks for a selection of trackways and/or populations across the *Australopithecus* to *Homo* transition. (**a**) Mean for Harry's Trackway in Namibia (see Sect. 1.4). (**b**) Mean for a hundred randomly selected tracks from the Bournemouth data (see Sect. 1.4). (**c**) Mean track for the G-1 Trackway at Laetoli (N = 11). (**d**) Mean track for the tracks at Ileret present on the upper surface (Fig. 3.6) (Modified from Morse et al. 2013)

Early African *Homo erectus* (Bennett et al. 2009) raised the possibility of looking more closely at the transition in foot-mechanics across the *Australopithecus* to *Homo* transition, although it is worth noting that Dingwall et al. (2013) do not discount the possibility that the tracks were made by a male *Paranthropus boisei*. Also Tuttle (1987) raised the idea that the Laetoli tracks may have been made by made by an as yet undiscovered hominin with greater similarity to *Homo* than *Australopithecus afarensis*. The latter species remains, however, the only hominin confirmed at Laetoli when the Footprint Tuff was deposited. Notwithstanding these caveats the question remains whether we can assemble and analyse as objectively as possible this track data using the tools and techniques explored in this book.

Figure 6.13 is an initial attempt to do exactly that and shows a comparison of four mean tracks: (1) a mean of 100 modern habitually shod humans based on the

Bournemouth data (see Sect. 1.4); (2) a mean track for Harry's Trackway in Namibia (Bates et al. 2013a; Morse et al. 2013); (3) a mean of 12 tracks from the Upper Footprint Surface at Ileret (Bennett et al. 2009; see Sect. 3.1.2); and (4) a mean of 11 tracks from the G-1 Trail at Laetoli (Crompton et al. 2012). The tracks are formed in different depositional environments and are likely to have had different bearing capacities. Those from Namibia and Ileret are both formed in fluvial over-bank deposits, compared to a laboratory sand tray (modern prints) and volcanic ash in the case of the Laetoli tracks. The effects of substrate as outlined in the models discussed in Sect. 5.3 (Figs. 5.11 and 5.14) are clearly visible in the Ileret mean; note the narrow heel of a deeper track and the longitudinal asymmetry with a deeper ball area compared to the heel. The degree to which such effects obscure the biomechanical signatures of the respective track-makers is open to debate. One interpretation is that the means show a remarkable level of consistency if one takes into account these substrate controls. For example, as one would expect the modern Bournemouth (habitually shod) and Namibian tracks (presumably Harry was habitually unshod) show few areas of difference as one would expect; the latter having a slightly flatter longitudinal arch typical of habitually unshod individuals. The Ileret data differs from the modern (Western + Namibian) tracks in having a deeper medial arch and deeper toe impressions due to the longitudinal asymmetry; while the Laetoli mean differs from the modern mean in having a deeper medial longitudinal arch (i.e. less prominent). The latter is also more restricted in extent, shallower hallucal impressions and a slightly deeper anterior heel impression. The Ileret and Laetoli means differ significantly only with respect to the deeper impression under the hallucal ray (the metatarsal head and distal phalanx or phalanges) in the case of the Ileret tracks. In summary taking into account the weaker substrate in the case of the Ileret tracks the difference between the means is potentially quite small.

Caution is clearly needed here because we are talking about limited track samples representing whole species with no idea of the typical distribution of topologies within them! How do we know that the Ileret or Laetoli tracks are for example representative of the species of track-maker that made them? Of course we don't which is why understanding the degree of variability within our own species is identified in the final chapter as a key research objective for the future. But with caution set aside for one moment, the results presented here could be interpreted as indicative of homeostasis in foot morphology across the *Australopithecus* to *Homo* transition consistent with the conclusions of Crompton et al. (2012) discussed in the previous section. Previous analysis by Bennett et al. (2009) suggested that the Ileret tracks showed a greater mechanical affinity to modern *Homo sapiens* tracks rather than to Laetoli. This work used a landmark based analysis and suggested that essentially modern lateral-to-medial force transfer across the metatarsal heads was present in the Ileret tracks, but less well developed in those from Laetoli. They also suggested that compared to Laetoli, the Ileret tracks have a more contracted proximal mid-foot region, including a deeper instep, suggesting the presence of better developed medial longitudinal arch. The data in Fig. 6.13 based on a 'whole foot' analysis refines this suggesting that the overlap in print topology may be greater and by inference that the foot morphology and function across the *Australopithecus* to *Homo* transition may not be as pronounced as first thought. Much of the observed differences can in fact be accounted for in terms of the influence of substrate. It is a conclusion that needs to be rigorously tested and the application of a whole-foot statistical comparison is now called for, but it remains an interesting assertion none the less.

6.6 Summary

Relationships between track dimensions and body proportions have the potential to allow the geoarchaeologist or palaeoanthropologist to make inferences on the stature and body mass of a track-maker, to infer the speed at which a series of tracks were made and perhaps to say something about the degree to which they conform, or diverge from, a stereotypical mode of locomotion. All of these inferences rely to varying degrees on reference to empirical models derived for comparative populations. Ensuring that appropriate models are used in making inferences is critical to the veracity of these inferences and few studies currently acknowledge the implicit assumptions that are made in using them. We draw attention to a number of potential issues here which need to be acknowledged. Firstly the degree to which particular empirical relationships, for example between stature and foot length, are representative of anything other than the individuals sampled. Crucially, is the sample representative of a wider population or not and is it really of sufficient size to be representative of the variance within that wider population? Many samples are limited in size and either consciously or unconsciously select from a restricted part of a wider population, for example the use of students, military personal or in the case of the Bournemouth data (see Sect. 1.4) university administrators and academic staff. We have demonstrated with respect to the Daasanach how similar sized samples from the same population give different results (cf. Bennett et al. 2009; Dingwall et al. 2013). Secondly, the degree to which a population (or in truth, sample) is representative of the likely track-maker, especially when dealing with extinct hominins, groups with different life styles and habits or where there is little contextual information. Thirdly the degree to which the tracks themselves may be a biased sample of a wider population, for example if substrate properties leads to the preferentially sampling of tracks made only by certain individuals (Falkingham et al. 2011), or if only selected parts of a population engage in an activity where their tracks may be preserved (Kinahan 2013). The more specific an empirical relationship the more chance there is of an error being made in its application. Generic multipliers such as those developed by Toppinard (1877) have appeal in light of this because they do not mislead by seducing the user by promise of accuracy and precision, they are exactly what they are, a general statement and nothing more. In many cases these issues are not crucial since the aim is to provide some descriptive colour, but in the case of extinct hominins where biometrics are important to help deduce the track-maker, or in forensic cases they are.

References

Adams BJ, Herrmann NP (2009) Estimation of living stature from selected anthropometric (soft tissue) measurements: applications for forensic anthropology. J Forensic Sci 54:753–760

Agnihotri AK, Purwar B, Googoolye K et al (2007) Estimation of stature by foot length. J Forensic Legal Med 14:279–283

Alexander RMN (1976) Estimates of speeds of dinosaurs. Nature 261:129–130. doi:10.1038/261129a0

Alexander RMCN (1984) Stride length and speed for adults, children, and fossil hominids. Am J Phys Anthropol 63:23–27

Anderson M, Blais M, Green WT (1956) Growth of the normal foot during childhood and adolescence: Length of the foot and interrelations of foot, stature, and lower extremity as seen in serial records of children between 1–18 years of age. Am J Phys Anthropol 14(2):287–308

Anton SC (2003) Natural history of *Homo erectus*. Yrbk Phys Anthropol 46:126–170

Ashizawa K, Kumakura C, Kusumoto A et al (1997) Relative foot size and shape to general body size in Javanese, Filipinas and Japanese with special reference to habitual footwear types. Ann Hum Biol 24(2):117–129

Ashton N, Lewis SG, De Groote I et al (2014) Hominin Footprints from Early Pleistocene Deposits at Happisburgh, UK. PLoS One 9(2):e88329

Atamturk D (2010) Estimation of sex from the dimensions of foot, footprints, and shoe. Anthropol Anz J Biol Clin Anthropol 68(1):21–29

Atamturk D, Duyar I (2008) Age-Related Factors in the Relationship Between Foot Measurements and Living Stature and Body Weight. J Forensic Sci 53(6):1296–1300

Avanzini M, Mietto P, Panarello A et al (2008) The Devil's Trails: Middle Pleistocene human footprints preserved in a volcanoclastic deposit of Southern Italy. Ichnos 15:179–189

Barker SL, Scheuer JL (1998) Predictive value of human footprints in a forensic context. Med Sci Law 38(4):341–346

Bassey EJ (1986) Demi-span as a measure of skeletal size. Ann Hum Biol 13(5):499–502

Bates KT et al (2013a) Does footprint depth correlate with foot motion and pressure? J R Soc Interface 10(83):20130009. doi:10.1098/rsif.2013.0009

Bates KT et al (2013b) The evolution of compliance in the human lateral mid-foot. Proc R Soc B Biol Sci 280:1769

Bennett MR, Harris JWK, Richmond BG et al (2009) Early Hominin Foot Morphology Based on 1.5 Million Year Old Footprints from Ileret, Kenya. Science 323:1197–1201

Bennett MR, Morse SA, Liutkus-Pierce C et al (2014) Exceptional preservation of children's footprints from a Holocene footprint site in Namibia. J Afr Earth Sci 97:331–341

Berge C, Penin X, Pellé É (2006) New interpretation of Laetoli footprints using an experimental approach and Procrustes analysis: Preliminary results. Comptes Rendus Palevol 5:561–569

Bertsch C, Unger H, Winkelmann W et al (2004) Evaluation of early walking patterns from plantar pressure distribution measurements. First year results of 42 children. Gait Posture 19(3):235–242

Bosch K, Gerß J, Rosenbaum D (2010) Development of healthy children's feet-Nine-year results of a longitudinal investigation of plantar loading patterns. Gait Posture 32:564–571

Bramble DM, Lieberman DE (2004) Endurance running and the evolution of Homo. Nature 432(7015):345–352

Brown T (1999) The science and art of tracking – nature's path to spiritual discovery. Berkley Books, New York

Charteris J, Wall JC, Nottrodt JW (1981) Functional reconstruction of gait from the Pliocene hominid footprints at Laetoli, northern Tanzania. Nature 290:496–498

Charteris J, Wall JC, Nottrodt JW (1982) Pliocene hominid gait: new interpretations based on available footprint data from Laetoli. Am J Phys Anthropol 58(2):133–144

Chhibber SR, Singh I (1970) Asymmetry in muscle weight and one-sided dominance in the human lower limbs. J Anat 106(3):553

Clarke RJ (1999) Discovery of complete arm and hand of the 3.3 million-year-old Australopithecus skeleton from Sterkfontein. S Afr J Sci 95:477–480
Crompton RH, Pataky TC, Savage R et al (2012) Human-like external function of the foot, and fully upright gait, confirmed in the 3.66 million year old Laetoli hominin footprints by topographic statistics, experimental footprint-formation and computer simulation. J R Soc Interface 9:707–719
Dangour AD (2003) Cross-sectional changes in anthropometric variables among Wapishana and Patamona Amerindian adults. Hum Biol 75(2):227–240
Davenport CB (1932) The growth of the human foot. Am J Phys Anthropol 17(2):167–211
Day MH, Wickens EH (1980) Laetoli pliocene hominid footprints and bipedalism. Nature 286:385–387. doi:10.1038/286385a0
Deloison Y (1991) Les australopitheques marchaient – ils comme nous? In: Coppens Y, Senut B (eds) Origine(s) de la Bipédie Chez les Hominidés. Editions du CNRS, Paris
Deloison Y (1992) Empreintes de pas à Laetoli (Tanzanie). Leur apport à une meilleure connaissance de la locomotion des Hominidés fossiles. Comptes rendus de l'Académie des sciences, Série 2, Mécanique, Physique, Chimie, Sciences de l'univers. Sciences de la Terre 315(1):103–109
Dingwall HL, Hatala KG, Wunderlich RE et al (2013) Hominin stature, body mass, and walking speed estimates based on 1.5 million-year-old fossil footprints at Ileret, Kenya. J Hum Evol 64(6):556–568
Dowling AM, Steele JR, Baur LA (2001) Does obesity influence foot structure and plantar pressure patterns in prepubescent children? Int J Obes Relat Metab Disord 25(6):845–852
Falkingham PL, Bates KT, Margetts L et al (2011) The 'Goldilocks' effect: preservation bias in vertebrate track assemblages. J R Soc Interface 8:1142–1154
Fawzy IA, Kamal NN (2010) Stature and body weight estimation from various footprint measurements among Egyptian population. J Forensic Sci 55(4):884–888
Friedlaender JS, Costa PT Jr, Bosse R et al (1977) Longitudinal physique changes among healthy white veterans at Boston. Hum Biol 49(4):541–558
Giles E, Vallandigham PH (1991) Height estimation from foot and shoeprint length. J Forensic Sci 36(4):1134–1151
Gordon CC, Buiskstra JE (1992) Linear models for the prediction of stature from foot and boot dimensions. J Forensic Sci 37(3):771–782
Grivas TB, Mihas C, Arapaki A et al (2008) Correlation of foot length with height and weight in school age children. J Forensic Legal Med 15:89–95
Hairunnisa MAK, Nataraja Moorthy T (2014) Stature estimation from the anthropometric measurements of footprint in Iban ethnics of east Malaysia by regression analysis. J Forensic Sci Criminol 1(5):1
Harcourt-Smith WE (2005) Did Australopithecus afarensis make the Laetoli footprint trail?-New insights into an old problem. Am J Phys Anthropol Suppl S40:116
Hemy N, Flavel A, Ishak NI et al (2013a) Sex estimation using anthropometry of feet and footprints in a Western Australian population. Forensic Sci Int 231(1):402-e1
Hemy N, Flavel A, Ishak NI, Franklin D (2013b) Estimation of stature using anthropometry of feet and footprints in a Western Australian population. J Forensic Legal Med 20(5):435–441
Hill LM (1958) Changes in the proportions of the female foot during growth. Am J Phys Anthropol 349–366
Hrdlička A (1935) The Pueblos. Am J Phys Anthropol 20:235–460
Ishigaki S, Lockley MG (2010) Didactyl, tridactyl and tetradactyl theropod trackways from the Lower Jurassic of Morocco: evidence of limping, labouring and other irregular gaits. Hist Biol 22(1–3):100–108
Jasuja OP, Singh J, Jain M (1991) Estimation of stature from foot and shoe measurements by multiplication factors: a revised attempt. Forensic Sci Int 52(2):203–215
Jowaheer V, Agnihotri AK (2011) Sex identification on the basis of hand and foot measurements in Indo-Mauritian population–A model based approach. J Forensic Legal Med 18(4):173–176

Kanchan T, Menezes RG, Moudgil R et al (2008) Stature estimation from foot dimensions. Forensic Sci Int 179(2):241-e1
Kanchan T, Krishan K, Sharma A et al (2010a) A study of correlation of hand and foot dimensions for personal identification in mass disasters. Forensic Sci Int 199(1):112-e1
Kanchan T, Menezes RG, Moudgil R et al (2010b) Stature estimation from foot length using universal regression formula in a North Indian population. J Forensic Sci 55(1):163–166
Kanchan T, Krishan K, Shyamsundar S et al (2012) Analysis of footprint and its parts for stature estimation in Indian population. Foot 22(3):175–180
Katzmarzyk PT, Leonard WR (1998) Climatic influences on human body size and proportions: ecological adaptations and secular trends. Am J Phys Anthropol 106(4):483–503
Kimbel WH, Delezene LK (2009) "Lucy" redux: A review of research on Australopithecus afarensis. Am J Phys Anthropol 140(S49):2–48
Kinahan J (2013) The use of skeletal and complementary evidence to estimate human stature and identify the presence of women in the recent archaeological record of the Namib desert. South Afr Archaeol Bull 68(197):72–78
Ko M, Hughes L, Lewis H (2012) Walking speed and peak plantar pressure distribution during barefoot walking in persons with diabetes. Physiother Res Int 17(1):29–35
Krishan K (2008a) Determination of stature from Foot and its segments in a north Indian population. Am J Forensic Med Pathol 29:297–303
Krishan K (2008b) Estimation of stature from Foot prints and Foot outline dimension in Gujjar of north India. Forensic Sci Int 175:93–101
Krishan K (2008c) Establishing correlation of footprints with body weight-Forensic Aspects. Forensic Sci Int 179:63–69
Krishan K, Sharma A (2007) Estimation of stature from dimensions of hands and feet in a North Indian population. J Forensic Legal Med 14(6):327–332
Krishan K, Kanchan T, Sharma A (2011a) Sex determination from hand and foot dimensions in a North Indian population. J Forensic Sci 56(2):453–459
Krishan K, Kanchan T, Passi N (2011b) Estimation of stature from the foot and its segments in a sub-adult female population of North India. J Foot Ankle Res 4(1):24
Kulthanan T, Techakampuch S, Donphongam N (2004) A study of footprints in athletes and non-athletic people. J Med Assoc Thailand 87(7):788–793
Lai PP, Leung AK, Li AN et al (2008) Three-dimensional gait analysis of obese adults. Clin Biomech 23:S2–S6
Leakey MD, Hay RL (1979) Pliocene footprints in the Laetoli beds at Laetoli, northern Tanzania. Nature 278:317
Lockley M, Roberts G, Kim JY (2008) In the Footprints of Our Ancestors: An Overview of the Hominid Track Record. Ichnos 15:106–125
Lovejoy CO (1988) Evolution of human walking. Sci Am 259(5):118–125
Macdonell WR (1902) On criminal anthropometry and the identification of criminals. Biometrika 1(2):177–227
Malina RM, Hamill PVV, Lemeshow S (1973) Selected body: measurements of children 6–11 years. DHEW Publication No. (HSM) 73-1605
Manna I, Pradhan D, Ghosh S, Kar SK et al (2001) A comparative study of foot dimension between adult male and female and evaluation of foot hazards due to using of footwear. J Phys Anthropol Appl Hum Sci 20(4):241–246
McAllister PJ (2011) The evolution of the inadequate modern male. Aust Sci May:19–21
McGraw B, McClenagham BA, Williams HG et al (2000) Gait and postural stability in obese and nonobese prepubertal boys. Arch Phys Med Rehab 81:484–489
Meldrum DJ (2000) Fossilized footprints in the Ka'u Desert, Hawaii. Am J Phys Anthropol Suppl 30:226–227
Meldrum DJ (2002) Midfoot flexibility and the evolution of bipedalism. Am J Phys Anthropol Suppl 34:111–112
Meldrum DJ (2004a) Fossilized Hawaiian footprints compared with Laetoli hominid footprints. In: Meldrum DJ (ed) From Biped to Strider. Springer, US, p 63–83

Meldrum DJ (2004b) Midfoot flexibility, fossil footprints, and sasquatch steps: New perspectives on the evolution of bipedalism. J Sci Explor 18(1):65–79
Meldrum DJ, Chapman RE (2007) Morphometrics of the outline shape of hominid footprints. Am J Phys Anthropol 42:170
Meldrum DJ, Lockley MG, Lucas SG, Musiba C (2011) Ichnotaxonomy of the Laetoli trackways: the earliest hominin footprints. J Afr Earth Sci 60(1):1–12
Mignardot JB, Olivier I, Promayon E et al (2010) Obesity impact on the attentional cost for controlling posture. PLoS One 5(12):e14387
Mohanty SP, Babu SS, Nair NS (2001) The use of arm span as a predictor of height: a study of South Indian women. J Orthop Surg 9(1):19–23
Morse SA, Bennett MR, Liutkus-Pierce C et al (2013) Holocene footprints in Namibia: the influence of substrate on footprint variability. Am J Phys Anthropol 151:265–279
Moudgil R, Kaur R, Menezes RG et al (2008) Foot index: Is it a tool for sex determination? J Forensic Legal Med 15(4):223–226
Muller S, Carlsohn A, Muller J et al (2012) Static and dynamic foot characteristics in children aged 1–13 years: A cross-sectional study. Gait Posture 35:389–394
Musiba CM, Tuttle RH, Hallgrímsson B (1997) Swift and sure-footed on the Savanna: A study of Hadzabe gaits and feet in Northern Tanzania. Am J Hum Biol 9(3):303–321
Mysorekar VR, Nandedkar AN, Sarma TCSR (1982) Estimation of stature from parts of humerus and radius. Med Sci Law 22(3):178–180
Nataraja Moorthy T, Wan NZ, Saat M (2011) A study on footprints of Malaysian athletes and non-athletes for application during forensic comparison. Malays J Forensic Sci 2:29–35
Nataraja Moorthy T, Ling AY, Sarippudin SA et al (2013a) Estimation of stature from footprint and foot outline measurements in Malaysian Chinese. Aust J Forensic Sci 1–24. doi:10.1080/00450618.2013.825813
Nataraja Moorthy T, MohdKhalil NAFBT, Khan HMA (2013b) Stature estimation based on footprint measurements of Malays in peninsular Malaysia by regression analysis. Int J Biomed Adv Res 4(10):683–689
Nataraja Moorthy T, Mostapa AMB, Boominathan R, Raman N (2014) Stature estimation from footprint measurements in Indian Tamils by regression analysis. Egypt J Forensic Sci 4(1):7–16
Ozden H, Balci Y, Demirüstü C et al (2005) Stature and sex estimate using foot and shoe dimensions. Forensic Sci Int 147(2):181–184
Pales L (1976) Les empreintes de pieds Humains dans les cavernes. Archives de l' Institut de Paléontologie Humaine 36:1–166
Pataky TC, Caravaggi P, Savage R et al (2008) New insights into the plantar pressure correlates of walking speed using pedobarographic statistical parametric mapping (pSPM). J Biomech 41(9):1987–1994
Pawar RM, Pawar MN (2012) Foot length – a functional parameter for assessment of height. Foot 22(1):31–34
Perissinotto E, Pisent C, Sergi G et al (2002) Anthropometric measurements in the elderly: age and gender differences. Br J Nutr 87(02):177–186
Raichlen DA, Pontzer H, Sockol MD (2008) The Laetoli footprints and early hominin locomotor kinematics. J Hum Evol 54:112–117
Raichlen DA, Gordon AD, Harcourt-Smith WEH et al (2011) Laetoli footprints preserves first direct evidence of human-like bipedal biomechanics. Plos One 5(3):e9769. doi:10.1371/journal.pone.0009769
Rao NG, Kotian MS (1990) Foot print ratio (FPR)-a clue for establishing sex identity. J Indian Acad Forensic Med 12(2):51–56
Reel S, Rouse S, Vernon W et al (2010) Reliability of a two-dimensional footprint measurement approach. Sci Justice 50:113–118
Reel S, Rouse S, Vernon W et al (2012) Estimation of stature from static and dynamic footprints Forensic Sci Int 219(1):283.e1–283.e5

Reynolds TR (1987) Stride length and its determinants in humans, early hominids, primates, and mammals. Am J Phys Anthropol 72(1):101–115

Robbins LM (1985) Footprints: collection, analysis, and interpretation. CC Thomas, Springfield

Robbins LM (1986) Estimating height and weight from size of footprints. J Forensic Sci 31(1):143–152

Roberts DF (1953) Body weight, race and climate. Am J Phys Anthropol 11(4):533–558

Roberts G, Gonzalez S, Huddart D (1996) Intertidal Holocene footprints and their archaeological significance. Antiquity 70:647–651

Roche AF (1986) Bone growth and maturation. In: Falkner F, Tanner JM (eds) Postnatal growth neurobiology. Springer, New York

Roche AF, Davila GH (1972) Late adolescent growth in stature. Pediatrics 50(6):874–880

Rosenbaum D, Hautmann S, Gold M et al (1994) Effects of walking speed on plantar pressure patterns and hindfoot angular motion. Gait Posture 2(3):191–197

Ruff C (2002) Variation in human body size and shape. Ann Rev Anthropol 31:211–232

Ruff CB, Walker A (1993) Body size and body shape. In: Walker AC, Leakey REF (eds) The Nariokotome *Homo erectus* Skeleton. Harvard University Press, Cambridge, MA, pp 234–265

Ruiz J, Torices A (2013) Humans Running at Stadiums and Beaches and the Accuracy of Speed Estimations from Fossil Trackways. Ichnos 20(1):31–35

Saltzman CL, Nawoczenski DA, Talbot KD (1995) Measurement of the medial longitudinal arch. Arch Phys Med Rehabil 76(1):45–49

Saxena SK (1984) A study of correlations and estimation of stature from hand length, hand breadth and sole length. Anthropol Anz: Bericht uber die biologisch-anthropologische Literatur 42(4):271–276

Schmid P (2004) Functional interpretation of the Laetoli footprints. In: Meldrum DJ (ed) From biped to strider. Springer, New York

Schmincke HU, Rausch J, Kutterolf S et al (2010) Walking through volcanic mud: the 2,100 year-old Acahualinca footprints (Nicaragua) II: the Acahualinca people, environmental conditions and motivation. Int J Earth Sci 99(1):279–292

Segal A, Rohr E, Orendurff M et al (2004) The effect of walking speed on peak plantar pressure. Foot Ankle Int 25(12):926–933

Sellers WI, Cain GM, Wang W et al (2005) Stride lengths, speed and energy costs in walking of Australopithecus afarensis: using evolutionary robotics to predict locomotion of early human ancestors. J R Soc Interface 2(5):431–441

Sellers WI, Manning PL (2007) Estimating dinosaur maximum running speeds using evolutionary robotics. Proc R Soc B 274:2711–2716

Sen J, Ghosh S (2008) Estimation of stature from foot length and foot breadth among the Rajbanshi: An indigenous population of North Bengal. Forensic Sci Int 181(1):55-e1

Sen J, Kanchan T, Ghosh S (2011) Sex estimation from foot dimensions in an indigenous Indian population. J Forensic Sci 56(s1):S148–S153

Singh TS, Phookan MN (1993) Stature and footsize in four Thai communities of Assam. India Anthropol Anz: Bericht uber die biologisch-anthropologische Literatur 51(4):349–355

Stavlas P, Grivas TB, Michas C et al (2005) The evolution of foot morphology in children between 6 and 17 years of age: a cross-sectional study based on footprints in a Mediterranean population. J Foot Ankle Surg 44:424–428

Stern JT Jr, Susman RL (1983) The locomotor anatomy of Australopithecus afarensis. Am J Phys Anthropol 60(3):279–317

Susman RL, Stern JT Jr, Jungers WL (1984) Arboreality and bipedality in the Hadar hominids. Folia Primatol 43(2–3):113–156

Taylor AJ, Menz HB, Keenan AM (2004) The influence of walking speed on plantar pressure measurements using the two-step gait initiation protocol. Foot 14(1):49–55

Toppinard P (1877) L'Anthropologie. Reinwald, Paris

Tuttle RH (1987) Kinesiological inferences and evolutionary implications from Laetoli bipedal trails G-1, G-2/3, and A. In: Leakey MD, Harris JM (eds) Laetoli: a Pliocene site in northern Tanzania. Clarendon, Oxford, pp 60–65

Tuttle RH (2008) Footprint clues in hominid evolution and forensics: Lessons and limitations. Ichnos 15(3–4):158–165

Tuttle RH, Webb D, Weidl E et al (1990) Further progress on the Laetoli trails. J Archaeol Sci 17:347–362

Tuttle RH, Webb DM, Baksh M (1991) Laetoli toes and Australopithecus afarensis. Hum Evol 6(3):193–200

Tyagi AK, Rani M, Kohli A (2004) Sexing by foot index. J Forensic Med Toxicol 21(1):10–11

Uhrová P, Beňuš R, Masnicová S (2013) Stature estimation from various foot dimensions among Slovak population. J Forensic Sci 58(2):448–451

Ukoha UU, EgwuOgugua A, Emmanuel UK et al (2013) Estimation of Stature using footprints in an adult student population in Nigeria. Int J Biomed Adv Res 4(11):827–833

Vallois H (1931) Les Impreintes de Pieds Humains des Grottes Pre-historiques du Midi de la France. Palaeobiologica 4:79–98

Webb S (2007) Further research of the Willandra Lakes fossil footprint site, southeastern Australia. J Hum Evol 52(6):711–715

Webb S, Cupper ML, Robins R (2006) Pleistocene human footprints from the Willandra Lakes, southeastern Australia. J Hum Evol 50(4):405–413

White TD, Suwa G (1987) Hominid footprints at Laetoli: Facts and interpretations. Am J Phys Anthropol 72:485–514. doi:10.1002/ajpa.1330720409

Wilder HH, Wentworth B (1918) Personal identification: methods for the identification of individuals, living or dead. RG Badger, Boston

Wilson ML, Lundy JK (1994) Estimated living statures of dated Khoisan skeletons from the south-western coastal region of South Africa. South Afr Archaeol Bull 49:2–8

Zeybek G, Ergur I, Demiroglu Z (2008) Stature and gender estimation using foot measurements. Forensic Sci Int 181(1):54-e1

Chapter 7
Forensics Applications

Abstract Traces left by track-makers whether walking or running in bare feet or encased in foot wear have the potential to help place an individual at a crime scene or to help investigators work out the pattern and sequence of actions that took place. While in Europe few people may move about barefoot, in other parts of the world a significant proportion of people may still be habitually unshod either out of preference or socio-economic necessity. In this context much of what we have discussed in this book is relevant to the forensic investigator and in this chapter we apply this knowledge within a forensic context.

7.1 Crime Scenes

Footwear and/or foot impressions can provide important sources of evidence within a range of criminal investigations (Lucock 1967; Cassidy 1980; Bodziak 2000). According to Qamra et al. (1980) footprint evidence was used in a criminal case as early as 1888 and literary examples abound in crime novels; the preverbal 'footprint in the flower bed'. In theory an individual involved at a crime scene will leave foot or footwear impressions en route to, at, and while exiting from a scene. Not only has this the potential to allow events to be reconstructed where traces overlap systematically, but may also provide a link between a suspect and the scene (Cassidy 1980; Naples and Miller 2004). According to Bodziak (2000) the skill of the forensic investigator is to first anticipate, look-for and record this evidence while at a scene and then to evaluate it accurately, making inferences that may help profile a suspect or link them to that scene. Much of the evidence at a crime scene and the focus for forensic officers is around two-dimensional traces; impressions left by a foot tracking mud, blood or other bodily fluids, around a crime scene. Three-dimensional tracks are rare within indoor scenes but may potentially exist more commonly at outdoor ones and may allow a suspect to be tracked to and from a scene (Bodziak 2000; DiMaggio and Vernon 2011). There are cases such as that in Fig. 7.1a which represent hybrids neither two-dimensional, nor strictly conventional tracks. In addition to criminal applications there is also a body of literature which looks to the foot as a potential source of corroborative evidence for the identification of victims particularly those from horrific incidents where severed body parts are common (e.g., Krishan 2008c; Krishan et al. 2011). Here a foot may have enhanced chances of survival encased

M.R. Bennett, S.A. Morse, *Human Footprints: Fossilised Locomotion?*,
DOI 10.1007/978-3-319-08572-2_7

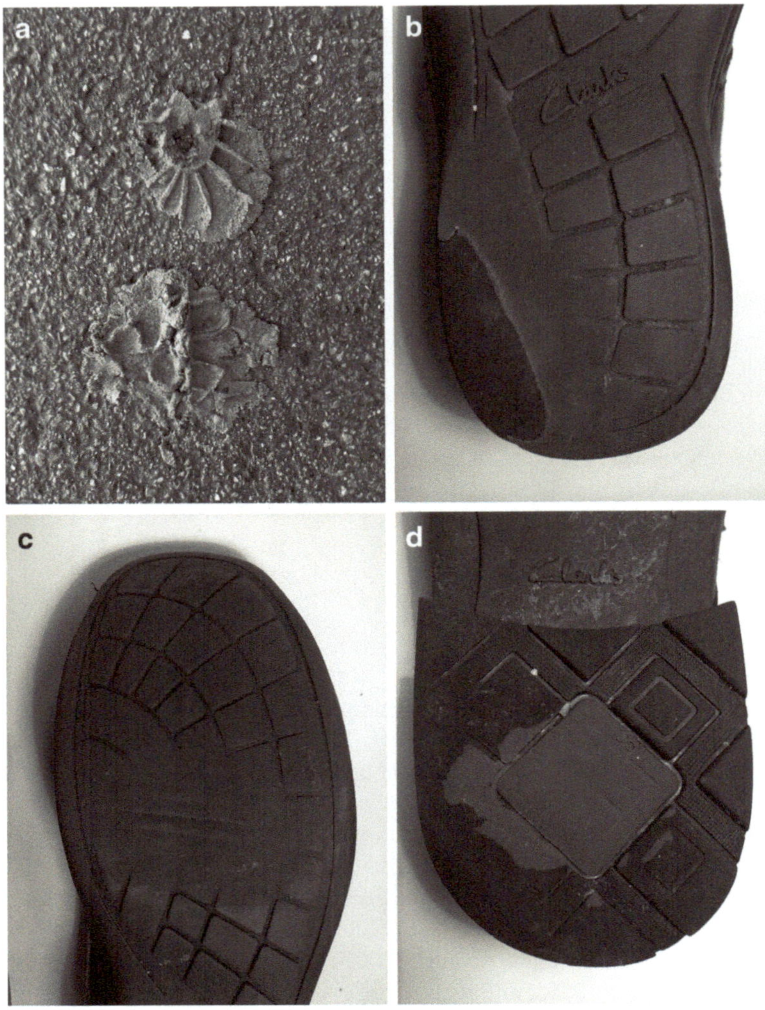

Fig. 7.1 Tracks and shoe wear patterns. (**a**) A hybrid track, neither strictly a two-dimensional nor a conventional three-dimensional track. In this case mud has been left as partial record of the passage of shoe on a pavement. (**b**–**d**) Wear patterns on shoes of the senior author. Note that none of these photographs are of forensic quality and are presented here simply for the purpose of illustration

within footwear. While this is not directly relevant to footwear and footprint evidence it is important that we acknowledge the potential of feet in this context.

We have already seen in the previous chapters how three-dimensional tracks can be recorded and used to infer such things as stature, body mass and gait of a track-maker. The question here is what can this essentially geoarchaeological perspective contribute to the investigation of a crime scene? The authors do not profess to be forensic experts and there are several definitive accounts on the evaluation of footwear traces to which the reader is referred (Cassidy 1980; Bodziak 2000; DiMaggio and Vernon 2011), but we do believe that the application of some of the geoarchaeological principles outlined here may be relevant to the forensic community or those practicing within it (Pringle et al. 2012). Our aim here is therefore to draw attention to those issues and to try not to stray too far into broader discussion of the footwear evidence in criminal investigations. In this context we recognise three broad areas where this might apply: (1) the collection of three-dimensional footwear and footprint evidence; (2) in evaluating the 'uniqueness' of the foot and gait; and (3) the degree to which it is possible to make inferences about a track-maker from the tracks they leave and the potential sources of concern that investigators need to be potentially mindful of.

7.2 Methods for Collecting Footwear Evidence

Footwear traces come in a variety of different forms and can be divided broadly into traces that are two-dimensional in which a residue is left and those that are three-dimensional impressions in which the substrate has been deformed in some way in a permanent or semi-permanent fashion. In practice in western society most outdoor evidence will involve shod individuals but this may not always be the case in large parts of Asia where for a variety of climatic and socio-economic factors many people still walk barefoot (Qamra et al. 1980; Krishan 2008a, b, c). Traditional methods and materials by which this trace evidence is collected are well documented in a number of crime scene manuals and have also grown in sophistication in recent years (Hueske 1991; Du Pasquier et al. 1996; Cassidy 1980; Bodziak 2000; Theeuwen et al. 2001; Buck et al. 2007). Our focus here is primarily with three-dimensional tracks that are collected and recorded in most cases currently using a combination of forensic photography and casting.

Bennett et al. (2009) suggested that traditional casting of footwear evidence while the norm is not without some issues. The first of these involves the very nature of the casting process which is by definition an invasive process with the potential to disturb the evidence that is being recorded if not conducted appropriately (Du Pasquier et al. 1996; Bodziak 2000). Essentially the investigator gets but one-shot to sample a series of tracks. The second issue involves the process of lifting a cast which will also remove trace evidence adhering to the basal surface of the cast and may obscure, at least initially, the footwear evidence preserved by the cast (Bull et al. 2006; Morgan et al. 2009). Clearly this trace evidence will have priority in any subsequent investigation and must be removed and analysed before the cast is available for study. The speed with which this is done depends on the resources available and the investigation priorities. There is an ever present risk that until the cast is cleaned it is not known whether the evidence has been successfully lifted and in many cases the evidence may have in the meantime been lost especially in transitory outdoor environments. The third issue involves the nature of the evidence itself; it is physical and bulky. Casts are bulky to store and not easily shared electronically between investigators. The net result is that tracks are primarily analysed in two-dimensions in the first instance and there is an increasing range of semi-automated systems that allow pattern matching within footwear databases and indexing with the aim of gathering intelligence (e.g., Davis 1981; Ashley 1996; Geradts and Keijzer 1996; Mikkonen et al. 1996; Alexander et al. 1999; Milne 2001; Napier 2002; Hannigan et al. 2006; AlGarni and Hamiane 2008; Pavlou and Allinson 2009). Three-dimensional data rarely forms a key part of this process, even where it exists and consequently potentially valuable data is being lost from the process especially with respect to quantifying wear patterns.

Bennett et al. (2009) argued that these issues could potentially be resolved by the application of three-dimensional data capture at a crime scene using the methods outlined in Sect. 2.4. By creating digital elevation models of a track at the crime scene the investigator has immediate access to three-dimensional data on footwear impressions which can be stored easily, shared and accessed electronically but crucially allows a more quantitative treatment with precise measurements being taken directly from a digital elevation model of a track. The collection of three-dimensional data by optical laser scanning has been demonstrated for the collection of footwear impressions in snow (Buck et al. 2007) and the documentation of soft tissue injuries (Thali et al. 2005) however there is still little work to date with respect to its application in the capture of footwear impressions. To demonstrate this point Bennett et al. (2009) conducted a number of experiments, which are reproduced here, using a tray (2.41 m long, 0.71 m wide and 0.09 m deep) filled with soft builder's sand with a moisture content of 2.88 % and a mean grain-size of 0.6 μm. In the first experiment a male subject (83 kg in weight, 1.81 m tall) walked the length of the sand tray wearing four different pairs of shoes creating a palimpsest of footwear impressions (US Size 9, UK Size 8.5, European Size 43, or 267 mm; Fig. 7.2). Two of these pairs of footwear were identical makes of boot, but with different degrees of wear. The sand tray was scanned and photographed with a Vi900 Konica-Minolta optical laser scanner and in addition the footwear used in the experiment was then mounted on an improvised cobbler's last and the soles scanned. In a second experiment a female subject (95 kg in weight, 1.53 m tall) walked the length of the sand tray wearing four different pairs of shoes creating a palimpsest of footwear impressions (US Size 7, UK Size 6, European Size 39, or 248 mm; Fig. 7.3).

The quality of the data capture is illustrated by the first of these experiments (Bennett et al. 2009; Fig. 7.2). All four shoes used in the experiment show varying degrees of wear characteristic of a slightly flat-footed individual. The two identical boots used in this experiment can be distinguished because of different degrees of wear on the lateral edge of the heel as well as the level to which cleats around the outside perimeter and circular traction domes on the sole have been removed by abrasion. There is no doubt that this could all be achieved by conventional forensic photography and casting, however the key advantage is the speed at which this can be accomplished using a single three-dimensional digital image which can be viewed from a variety of angles and with different levels of illumination. Multiple photographs and casts would be required to achieve the same by more conventional methods. The scans however have another significant advantage in that they allow the degree of wear to be quantified easily and accurately.

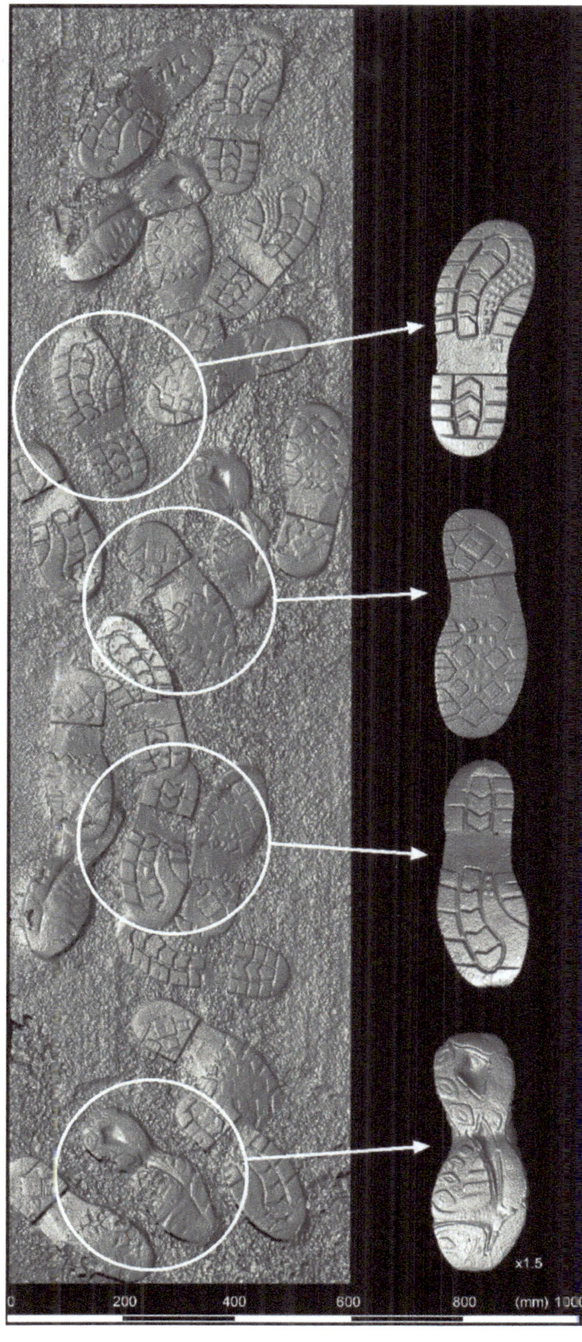

Fig. 7.2 Forensic experiment within a sand tray using two boots of the same size with different levels of wear (Modified from Bennett et al. (2009))

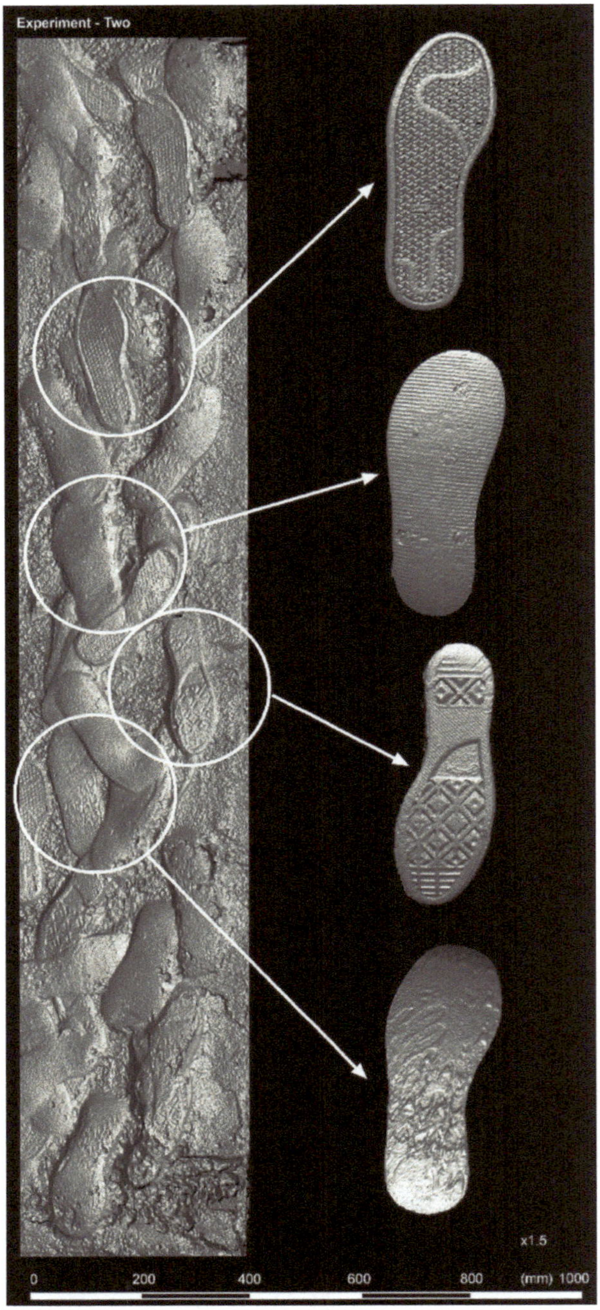

Fig. 7.3 Forensic experiment within a sand tray using a range of different shoes worn by the same track-maker (Modified from Bennett et al. (2009))

Again the second experiment (Fig. 7.3) illustrates the sheer quality and visual impact of the scans obtained and the ability to capture a large area not just the individual tracks. The different footwear is clearly distinguishable and all prints can be assigned easily to one of the four pairs of shoes using the tread patterns and sole outlines. Although this could be undertaken from vertical photographs, especially taken from varying angles and light levels, the advantage is that the single scan can be rotated, enlarged, rendered with different colours that mirror depth, and the degree of illumination and its angle changed to optimise recognition. It is also possible to measure key track dimensions to help verify the correct identification especially where two tracks are made by identical shoes but of different sizes. The advantages are clearly in the superior quality of the digital information and the ability to return to the 'virtual crime scene' repeatedly as an investigation develops.

While demonstrating potentially superior visualisation options these experiments are rather limited being laboratory based. Bennett et al. (2009) suggest that there are two challenges here. One is to demonstrate the value to the forensic community welded to an existing tool kit, while the second is to tackle the issue of operational deployment. Deploying an optical laser scanner is not always easy (Table 2.1). The relatively high capital costs, coupled with the need for a protective rig in which to house the scanner from ambient light and dust pose some significant challenges especially when operating in a confined setting. However as discussed in Sect. 2.4 the potential for digital photogrammetry to remove some of these challenge is emerging. The illustrations in Fig. 7.4 are all of tracks made by a Wellington Boot just above the high tide line on the Conwy Estuary in North Wales with a variety of seaweed and other algal vegetation covering the surface. Multiple oblique photographs were taken of the individual tracks and photogrammetric models made following the approach outlined by Falkingham (2012). They illustrate clearly the quality of the data that can be obtained with the right software. Photogrammetry provides therefore a very real opportunity for effective operational deployment. It is a realistic request for a crime scene photographer to capture 'additional and multiple oblique' images and for them to be processed subsequently providing in the first instance an additional source of information on three-dimensional tracks to complete existing approaches. The challenge at the moment is that the user interfaces for the freeware used by Falkingham (2012) is not ideal, although being developed rapidly. It is important to note that several proprietary software products are now selling their wares to the forensic community, although in the authors' experience they are much more limited in quality than they claim, but it is only a matter of time before these tools are developed further and become widely available. It is important to emphasise that in comparing the results from optical laser scanning and photogrammetry Bennett et al. (2013) found that the key limitation of photogrammetry was in the accurate scaling of the models produced and in a forensic context the need for independent testing and calibration of this scaling is essential.

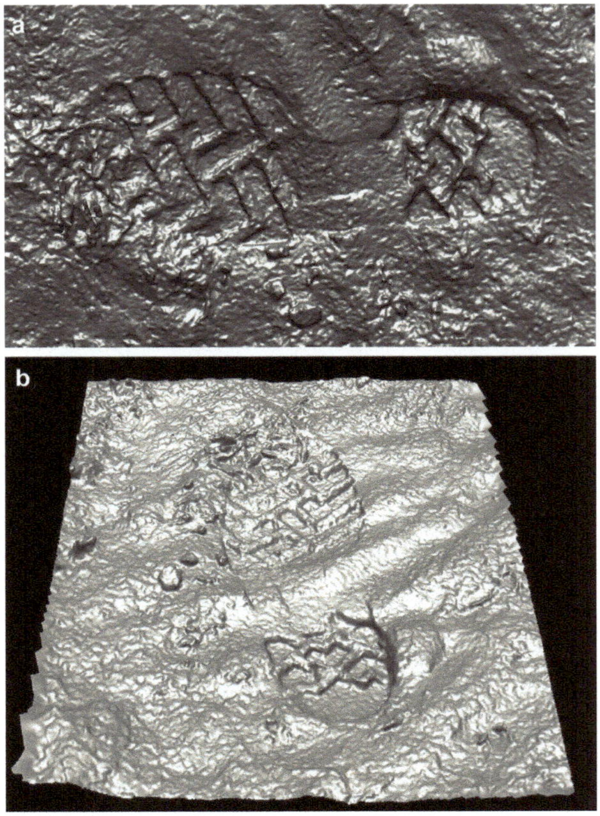

Fig. 7.4 Images of three-dimensional scans created using photogrammetry of a UK Size 9 Wellington Boot in estuarine muds

7.3 How Unique Is a Human Track?

It is widely accepted that one's fingerprint is to some degree unique, but can the same be said of one's footprint and pattern of gait? In a forensic context this is potentially an important question and while there is a growing body of empirical research to suggest that this might be true at least to some extent. The application of barefoot analysis in criminal cases is not without controversy (Tuttle 2008; Petraco et al. 2010). Moreover the whole concept of 'individualisation' or 'uniqueness' in forensic science is currently a topic of debate and crucially the way in which such evidence should be presented to avoid misleading criminal investigations and court cases (e.g., Stone 2006; Evett et al. 1998; Kerstholt et al. 2007; Saks and Faigman 2008; Cole 2009; Coyle et al. 2009; Koehler 2011). Notwithstanding these issues of probability, and their presentation in court, at the heart lies the idea that the human foot leaves a distinctive and potentially unique track. Under normal walking a human foot interacts with the ground in a stereotypical fashion taking on average 0.7 s to do so at a speed of 1.2 ms^{-1} (Pataky et al. 2012). The variable motion of the body and limbs during motion (Blanc et al. 1999) when coupled with natural variation in the anatomy of the human foot of the track-maker may introduce sufficient variance to allow a track to be distinct with at least higher orders of probability whether with respect to its outline, dimensions or with respect to pressure distribution. On top of this distinctiveness we may add those factors specific to an individual associated with such things as specific pathologies, deformities or injuries. The question that one must address is whether elements of uniqueness exist despite stereotypical footfall and the basic anatomical similarity of human feet?

A claim for uniqueness within human tracks has been made by some (Robbins 1978, 1985, 1986) and the potentially distinctive nature of tracks is widely recognised in the forensic literature (Sharma 1980; Qamra et al. 1980; Laskowski and Kyle 1988; Barker and Scheuer 1998; Bodziak 2000; Massey 2004; DiMaggio and Vernon 2011). There is a growing body of empirical research driven primarily by the work of the Royal Canadian Mounted Police to support this with respect to multiple foot dimensions (Kennedy 1996, Kennedy et al. 2003, 2005; Kennedy and Yamashita 2007; Yamashita 2007). Building on an earlier pilot study, Kennedy et al. (2005) presented a statistical framework based on just under 6,000 individual records drawn from what they describe as a 'general population' of mixed age, sex and race which suggested that the chances of a unique match between barefoot prints was the order of 7.88×10^{-10} or that there was one in 1.27 billion of two individuals producing the same outline. This was based on two-dimensional barefoot impressions collected on inkless paper and a total of up to 323 measurements (more typically c. 200) were made from each foot depending on the size and nature of a specific track using a semi-automated approach. The measurements fall broadly into five groups: (1) foot measurements such as lengths and widths; (2) F-points which are coordinates of point around the foot; (3) L-points which are width slices orthogonal to the axis of the foot defined by bisecting an enveloping cone formed by lines tangential to the inner and outer line of the foot (Fig. 2.9); (4) areas of such things as toe pads or the sole; and (5) finally a variety of angles between linear measurements. The model was cross-validated by records excluded from the original model and tested via multiple repetitions. This work is based on a robust body of data, with a sound statistical foundation and gives real strength to the assertion that barefoot tracks are to some extent unique to a specific track-maker (Kennedy et al. 2005). It is important to recognise that it is an empirical study however and that even though the sample on which it is based is large and drawn from a 'mixed population' it is still specific to that population and may not be applicable to other populations as discussed in Sect. 6.6. The size of the database of two-dimensional tracks collected by the Royal Canadian Mounted Police continues to grow and according to DiMaggio and Vernon (2011) it currently has over 24,000 records and growing. Yamashita (2007) discusses the widespread application of this work in court, although there have been some cases where this type of evidence has been refuted. To be clear their claim is not that barefoot impressions are unique, but that there is a high level of statistical probability to support matching (or not) of tracks thereby providing evidence to link (or not) a suspect to a crime scene where such tracks are found.

Table 7.1 Guidance for the evaluation of scientific expert witness testimony in light of US Supreme Court Rulings (Modified from Grivas and Komar (2008))

Guidelines from the Daubert decision	
1	Be testable and have been tested through scientific method
2	Have been subject to peer review
3	Have established methods
4	Have a known or potential error rate
5	Have widespread acceptance by the relevant scientific community
Guidance from the Kumho decision	
1	Expert witnesses can develop theories based on their observations and experience and then apply those theories to the case before the court
2	All forms of expert witness testimony should be evaluated with the same level of rigor
3	The Daubert standards are flexible guidelines that may not be applicable in every instance or expert witness testimony

In several wonderfully direct and acerbic pieces Tuttle (1986, 2008) challenges the issue around the use of barefoot evidence in court cases, recounting his encounters with the late Louise Robbins who provided evidence in a number of high profile court cases in the late 1970s and early 1980s. He made a number of clear and valid recommendations which still hold: (1) that any data and methods by self-proclaimed footprint experts need to be rigorously peer reviewed outside the court room before they enter it; (2) that the credentials of foot experts need to be certified and verified in some way as well as being limited to their area of expertise; and (3) all new forensic tools need to be subject to rigorous scientific testing before they are applied in criminal cases. These principles are not that different from the guidelines issued by the US Supreme Court in light of the *Daubert v. Merrill Dow Pharmaceuticals, Inc.* (1993) and while these have led to a range of legal challenges (so-called Daubert Motions), some negative comment (Grivas and Komar 2008) and may have impacted adversely in some civil cases (Berger 2005) they are sound scientific principles and remain so despite the modification in light of the *Kumho Tire, Ltd v. Carmichael* (1999) rulings (Table 7.1).

While the growing body of research, by such people as Kennedy and his colleagues, is clearly beginning to demonstrate a scientific foundation for footprint recognition more is needed and the limitations of such studies need to be clearly stated in their application. We have already established the 'empirical' nature of this type of evidence base and it is crucial that the appropriateness of the empirical model in a given case is established (see Sect. 6.6). It is no doubt easy to 'wow' a jury with a statistic like 'one in 1.27 billion' but how applicable is the sample and the population from which it is drawn to the case in hand? Coyle et al. (2009) make a number of important points about the miss-use and abuse of statistics in the court room drawing on the work of Kennedy et al. (2005) in illustration, pointing out that the operator error in collecting and initially analysing a track is far more significant, but often neglected, in presenting probability estimates to support assertions made in court. This also resonates with the wider debate in forensic science alluded to at the start around the question of what constitutes 'uniqueness'.

The uniqueness of a person's gait has been explored recently by Pataky et al. (2012). They used plantar pressure images for a 104 individuals to demonstrate that each has a potentially different pressure record. An average pressure distribution, through time during normal walking, was calculated for each individual with right and left feet being treated independently. Image processing and feature extraction was used to build a robust classification model that was successful in over 98 % of cases in discriminating one individual from the rest, suggesting a level of 'uniqueness' within the constraints of the sample. The study is not without limitations since plantar pressure records are known to vary with walking speed and other environmental factors (Rosenbaum et al. 1994, Rosenbaum and Becker 1997), but it does indicate that pressure records may be quite distinctive. Further validation of this is clearly required with greater sample sizes and the introduction of other variables, but is an intriguing study. One might suspect that if individuals have distinctive plantar pressure patterns then they should leave distinctive tracks that are in some way specific to them. We are careful here not to imply any measure of 'uniqueness' which would need to be established and statistically defined. Bates et al. (2013) demonstrated at least for shallow tracks a reasonably correlation between track depth and plantar pressure distributions following the observations of others (D'Aout et al. 2010). In Fig. 7.5 we have randomly selected a series of 24 male and female subjects from the Bournemouth data (see Sect. 1.4) and presented their mean track created using Pedobarographic Statistical Parametric Mapping (pSPM; see Sect. 2.7). The diversity of track topologies present given the uniform substrate, walking speeds and environmental conditions is quite striking and remains so if one looks at all 254 subjects in the data set despite the broad similarities associated with the stereotypical pattern of modern human locomotion. The sheer variety of track topologies is also illustrated by a review of the tracks presented in the Appendix. The degree to which these distributions are reproducible on multiple occasions by the same subjects and the degree to which they differ between individuals need to be explored and validated through further research, but it does reveal a glimpse of what might be possible given rigorous research and subsequent peer validation.

The idea that each of us has a distinctive gait underlies in part to the use of wear patterns on the outsoles of shoes to help link a shoe, and by association the owner, to a crime scene (Facey et al. 1992; Cassidy 1980; Bodziak 2000; DiMaggio and Vernon 2011). It is important to emphasis here that this is a multi-part challenge; first to link a shoe to a scene, second to demonstrate the ownership of that shoe by a suspect, and third that they in fact wore that shoe at the time in question. This association may be achieved and corroborated through such things as proof of purchase, witness statements, trace evidence and DNA but we draw attention here to the potential of shoe insoles to bear a distinct barefoot impression of the habitual shoe wearer (DiMaggio and Vernon 2011).

Evett et al. (1998) provides a robust framework in which to explore the issues of foot and footwear evidence and have more recently (Cook et al. 1998a, b; Evett et al. 2000) emphasised the importance of defining formal 'propositions', one for the prosecution and one for the defence, at the outset of an investigation to assist in the assessment, evaluation and presentation of such evidence. These

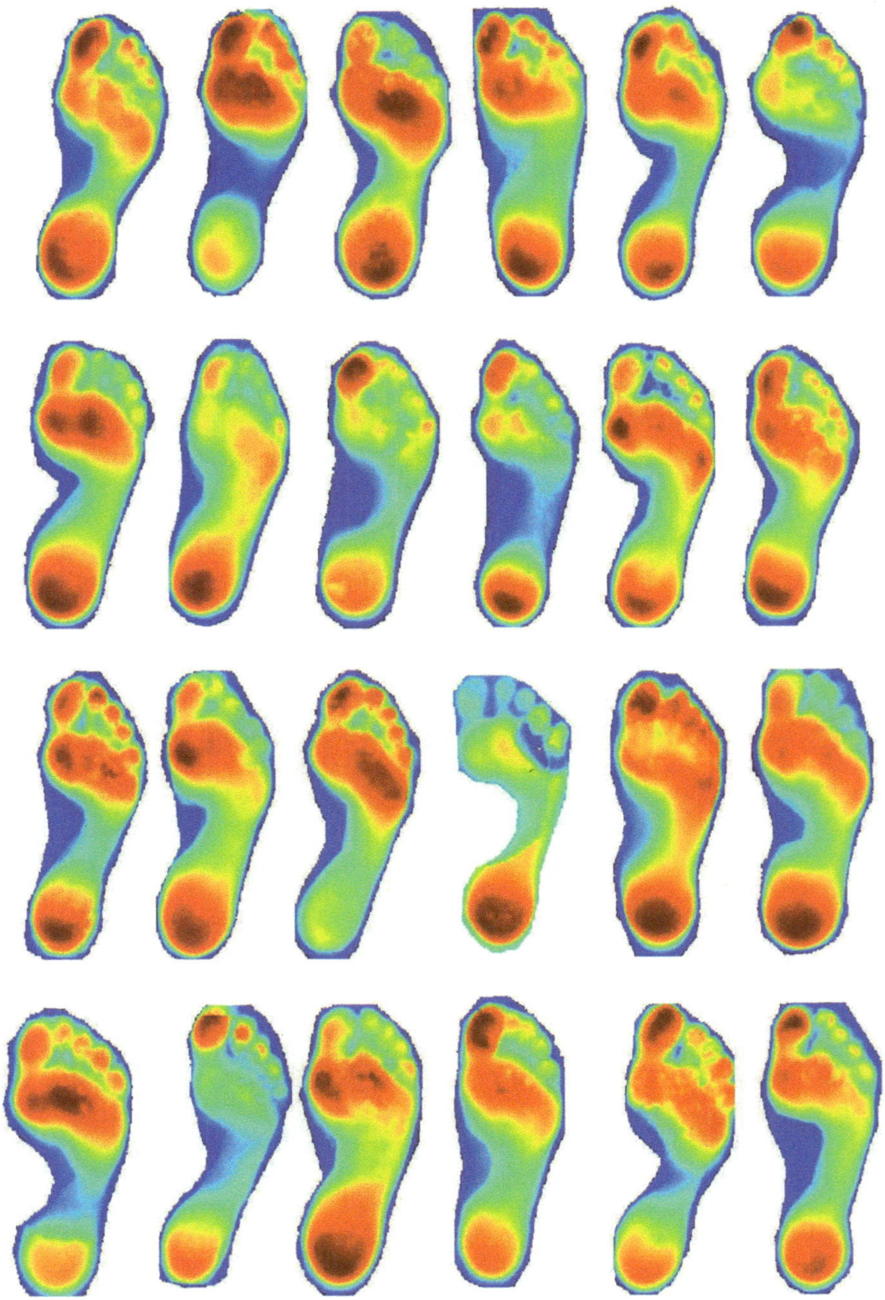

Fig. 7.5 A selection of 24 mean tracks taken at random from the 254 possible tracks within the Bournemouth data (see Sect. 1.4). Note the lack of apparent similarity between the individual tracks

propositions should be developed through informal 'explanations' to ensure that they are robust and focused on the right question. In this work they also make a clear distinction between 'identification' and 'individualisation'; the former while often used loosely is perhaps best restricted to the placement of an object into a restricted class, while the latter refers to defining the specific link between a trace and an individual and requires an assessment in some form of 'uniqueness' (Evett et al. 1998). This is helpful in considering a structured approach to the assessment of a series of tracks (Skerrett et al. 2011), whether they be made by shod or unshod individuals and allows one to frame a clear hierarchy of propositions. Starting with the highest order, we have:

Class characteristics: These allow identification, namely to establish the presence or absence of a particular foot or a shoe at a crime scene (Hancock et al. 2012). In the context of footwear this is about establishing the identity in terms of the make of the shoe and the size of a track at a crime scene and potentially then to say whether a suspect's shoe belongs to the same class (i.e., could it have made the track?). In the context of footwear they are largely the characteristics that result from the manufacture of a particular item of footwear. It is important to emphasise that there is nothing here about individualisation. There are a lot of shoes in the World, in fact around 13 billion pairs of shoes in 2005 (~6.6 billion population) of which 2.4 billion were owned in the US alone, across 6,000 different makes with 600 new ones appearing each year according to data cited in Smith (2009). This is a lot of shoes, but despite this one can narrow the class boundaries substantially since via size, design and crucially variations in the moulds used in sole manufacture (Bodziak 2000; Smith 2009). Again drawing on data cited in Smith (2009) the Nike 'Air Force I' first introduced in 1982 has sold well over 33 million pairs, but the model range involves over a 1,000 moulds each with a subtle variation in design; for example a size 8½ is associated with 75 different moulds alone. Law enforcement agencies hold a variety of two-dimensional databases which keep track of this diversity (Bodziak 2000; Smith 2009). In the context of bare feet, class characteristics are determined by aspects of shape and size (Qamra et al. 1980; Robbins 1985; Vernon 2006; DiMaggio and Vernon 2011). Essentially do the dimensions and outline of a suspect's foot match those of a track or trace? We would emphasise that it is important here that like is being compared with like; we have seen for example in Sect. 6.3 that different measurements can be obtained from a true three-dimensional track from one that is two-dimensional or derived from direct measurements.

Systematic identifying characteristics. If we accept the premise for one moment that each of us walks to some degree with a distinctive pattern of gait the wear characteristics of our shoes and of on feet themselves (i.e. callouses and hard skin) should reflect this. Consider the images in Fig. 7.1b–d which show a selection of the senior author's shoes. He walks to work each day, has a limited number of shoes, which he treats poorly and wears until they fall apart. He has a pattern of gait which leads to pronounced wear on the lateral edge of the heel and central portion of the ball, a pattern which is also reflected on his bare feet by the pattern of hardened skin. These shoes are highly distinctive and are likely to leave a clear record in both two- and three-dimensional tracks. The tracks shown in Fig. 7.2 made by two shoes with different wear are clearly distinct and the value of wear in distinguishing two dimensional tracks is well documented (e.g., Fruchtenicht et al. 2002; Adair et al. 2007). While these are perhaps more extreme examples the potential for repetitive contact between the sole of a shoe and the ground is well established (Bodziak 2000; Smith 2009). Abrasion patterns on the outsoles of shoes which typically form a series of microscopic ridges with an anastomosing form are referred to as Schallamach Patterns (Schallamach 1968) or feathering. Davis and Keeley's (2000) suggested that Schallamach Patterns developed within as little of 6 h of shoe wear and are highly distinctive, developing in different ways on identical outsoles. The limitation is that these abrasion patterns constantly evolve with continued wear and matching between a crime scene and a shoe needs to be achieved within a relatively short window. Despite this they have the potential to match a shoe or a trace and have been used to do so in criminal cases (Smith 2009). This style of abrasion pattern is only visible in two-dimensional traces being typically too fine to be preserved in three-dimensional tracks. In the case of barefoot impressions the outline of the foot, the presence of creases and humps (Qamra et al. 1980) and the length plus disposition of the toes can all be very distinctive and provide further refinement of 'class characteristics' as illustrated by a review of the tracks in the Appendix. The degree to which a distinctive plantar pressure pattern is recorded in the depth of a three-dimensional track may also lead to systematic track typologies, given comparative substrate properties and walking speeds (Fig. 7.5). The degree to which this holds true and is both consistent for an individual and distinctive between individuals needs to be explored further but may have the potential in the future to help refine class characteristics.

Random identifying characteristics. Our feet and shoes are all subject to random or semi-random (associated with habit and life style choices) damage which may leave marks on a shoe or directly on our feet in the form of scars, lesions and deformities (DiMaggio and Vernon 2011). In the case of footwear Stone (2006) provides a systematic way for recording this and also assessing the probability that a similar pattern of damage may occur by chance. This involves placing a grid of 16,000 1 mm squares across the shoe and using this to record the position, configuration and orientation of the damage. The probability of a match based on one characteristic is 1 in 16,000 rising rapidly to 1 in 127,992,000 for two occurrences and so on (Stone 2006). Other recording systems exist such as the focal point method of Vernon et al. (1999) which involves reference to 21 placed landmarks or regions on the sole. The application of spatial statistics to compare footwear marks is increasing (e.g., Sheets et al. 2013) and this is a potentially fertile area of further research, applying the techniques and principles of geometric morphometrics outlined in Sect. 2.7 to the quantitative comparison of accidental wear marks. In the context of barefoot impressions the linking of podiatry records to a foot may be particularly important (DiMaggio and Vernon 2011). In all these cases we are dealing with elements that allow a unique match to be made and therefore with the concept of individualisation.

Linking footwear and feet to a crime scene is therefore possible with care and robust analysis and we direct readers to the specialist texts of Cassidy (1980), Bodziak (2000) and DiMaggio and Vernon (2011).

7.4 Profiling a Suspect

In Sect. 6.3 we explored the plethora of empirical models available to infer an individual's body dimensions from the tracks they leave. In a forensic context being able to assist investigators with information to fill out a description of a potential suspect is appealing. Clearly this has the greatest potential in the case of barefoot tracks, although is not restricted to such cases alone (Bodziak 2000). Empirical relationships between shoe sizes and stature do exist (e.g., Giles and Vallandingham 1991) but this is complicated by the fact that a suspect may have a range of shoes that vary subtly in size depending on the style of the shoe or the availability of different sizes (Naples and Miller 2004). In Sect. 6.3 we discussed how these empirical models are extremely specific to the population on which they are based and vary with such things as sex, ethnicity/race and in some cases with bilateral asymmetry. As Fig. 6.5 demonstrates a wide range of stature estimates can be obtained for a given set of foot lengths depending on the model used and all rely on slightly different measures. Selecting the most appropriate model, assuming one exists, is critical. The advantage of a forensic, as opposed to geoarchaeological, scenario however is that there is greater potential for other contextual information to help constrain this selection, for example independent clues with respect to the likely sex and ethnicity/race of the suspect. Clearly the purpose of the profile is important here; if it is simply to add textual detail to a general suspect description based on multiple criteria then the risks of misdirection based on inferring stature from a track is likely to be low. However if it is to be used more precisely to, for example, rule a suspect in or out of an investigation then the intendant risks are much greater and we would argue that the assumptions and appropriateness of any empirical model applied need to be articulated and presented in a very transparent and open fashion. This also holds true with respect to the use of feet to help identify victims of natural disasters or terrorist atrocities through provision of corroborative information on such things as sex and stature (Krishan 2008a, b, c).

7.5 Summary

The application of some of the data collection and analytical tools discussed in this book within a forensic context is clear. The creation of digital elevation models of tracks for example has in the authors' opinion real potential to enhance the quality of information available to an investigator and is increasingly within the operational grasp of crime scene units give the application of photogrammetry. The degree to which human tracks are 'unique' in a forensic context is debatable but the work of the Royal Canadian Mounted Police suggests that foot outlines are distinctive to high levels of probability within the context of the mixed population they have sampled. There is no doubt that tracks whether made by shod or unshod suspects can contain information that may help link them (or not) to a particular crime scene and within the limitations of empirically based models may help provide data with which to profile suspects. There is a scope for much more research here, however, to underpin the usefulness of footwear evidence, especially three-dimensional evidence in a criminal context.

References

Adair TW, Lemay J, McDonald A et al (2007) The Mount Bierstadt Study: an experiment in unique damage formation in footwear. J Forensic Identif 57:199–205

Alexander A, Bouridane A, Crookes D (1999) Automatic classification and recognition of shoeprints. Image Process Appl Seventh Int Conf 465(2):638–641

AlGarni G, Hamiane M (2008) A novel technique for automatic shoeprint image retrieval. Forensic Sci Int 181:10–14

Ashley W (1996) What shoe was that? The use of computerised image database to assist in identification. Forensic Sci Int 82:7–20

Barker SL, Scheuer JL (1998) Predictive value of human footprints in a forensic context. Med Sci Law 38(4):341–346

Bates KT, Savage R, Pataky TC et al (2013) Does footprint depth correlate with foot motion and pressure? J R Soc Interface 10(83):20130009

Bennett MR, Huddart D, Gonzalez S (2009) Preservation and analysis of three-dimensional footwear evidence in soils: the application of optical laser scanning. In: Ritz K, Dawson L, Miller D (eds) Criminal and environmental soil forensics. Springer, Dordrecht, pp 445–461

Bennett MR, Falkingham P, Morse SA et al (2013) Preserving the impossible: conservation of soft-sediment hominin footprint sites and strategies for three-dimensional digital data capture. PLoS One 8(4):e60755. doi:10.1371/journal.pone.0060755

Berger MA (2005) What has a decade of Daubert wrought? Am J Public Health 95:s59–s65

Blanc Y, Balmer C, Landis T et al (1999) Temporal parameters and patterns of the foot roll over during walking: normative data for healthy adults. Gait Posture 10(2):97–108

Bodziak WJ (2000) Footwear impression evidence. CRC Press, Boca Raton

Buck U, Albertini N, Naether S et al (2007) 3D documentation of footwear impressions and tyre tracks in snow with high resolution optical surface scanning. Forensic Sci Int 171:157–164

Bull PA, Parker A, Morgan RM (2006) The forensic analysis of soils and sediment taken from the cast of a footprint. Forensic Sci Int 162:6–12

Cassidy MJ (1980) Footwear identification. Public Relations Branch of the Royal Canadian Mounted Police, Ottawa

Cole SA (2009) Forensics without uniqueness, conclusions without individualization: the new epistemology of forensic identification. Law Probab Risk 8(3):233–255

Cook R, Evett IW, Jackson G et al (1998a) A hierarchy of propositions: deciding which level to address in casework. Sci Justice 38:231–239

Cook R, Evett IW, Jackson G et al (1998b) A model for case assessment and interpretation. Sci Justice 38:151–156

Coyle IR, Field D, Wenderoth P (2009) Pattern recognition and forensic identification: the presumption of scientific accuracy and other falsehoods. Crim Law J 33:214–226

D'Aout K, Meert L, Van Gheluwe B et al (2010) Experimentally generated footprints in sand; analysis and consequences for the interpretation of fossil and forensic footprints. Am J Phys Anthropol 141:515–525

Davis RJ (1981) An intelligence approach to footwear marks and toolmarks. J Forensic Sci Soc 21:183–193
Davis RJ, Keeley A (2000) Feathering of footwear. Sci Justice 40:273–276
DiMaggio JA, Vernon W (2011) Forensic podiatry: principles and methods. Springer, Dordrecht
Du Pasquier E, Hebrard J, Margot P et al (1996) Evaluation and comparison of casting materials in forensic sciences applications to tool marks and foot/shoe impressions. Forensic Sci Int 82:33–43
Evett IW, Lambert JA, Buckleton JS (1998) A Bayesian approach to interpreting footwear marks in forensic casework. Sci Justice 38(4):241–247
Evett IW, Jackson G, Jones PJ et al (2000) More on the hierarchy of propositions: exploring the distinction between explanations and propositions. Sci Justice 40:3–10
Facey OE, Hannah ID, Rosen D (1992) Shoe wear patterns and pressure distribution under feet and shoes, determined by image analysis. J Forensic Sci Soc 32:15–25
Falkingham PL (2012) Acquisition of high resolution three-dimensional models using free, open-source, photogrammetric software. Palaeontol Electronica 15:1T:15p
Fruchtenicht TL, Herzig WP, Blackledge RD (2002) The discrimination of two-dimensional military boot impressions based on wear patterns. Sci Justice 42:97–104
Geradts Z, Keijzer J (1996) The image-database REBEZO for shoeprints with developments on automatic classification of shoe outsole designs. Forensic Sci Int 82:21–31
Giles E, Vallandigham PH (1991) Height estimation from foot and shoeprint length. J Forensic Sci 36(4):1134–1151
Grivas CR, Komar DA (2008) Kumho, Daubert, and the nature of scientific inquiry: implications for forensic anthropology. J Forensic Sci 53:771–776
Hancock S, Morgan-Smith R, Buckleton J (2012) The interpretation of shoeprint comparison class correspondences. Sci Justice 52:243–248
Hannigan TJ, Fleury LM, Reilly RB et al (2006) Survey of 1276 shoeprint impressions and development of an automatic shoeprint pattern matching facility. Sci Justice 46:79–89
Hueske E (1991) Photographing and casting footwear/tiretrack impressions in snow. J Forensic Identif 41:92–95
Kennedy RB (1996) Uniqueness of bare feet and its use as a possible means of identification. Forensic Sci Int 82(1):81–87
Kennedy RB, Yamashita AB (2007) Barefoot morphology comparisons: a summary. J Forensic Identif 57(3):383
Kennedy RB, Pressman IS, Chen S et al (2003) Statistical analysis of barefoot impressions. J Forensic Sci 48(1):55–63
Kennedy RB, Chen S, Pressman IS et al (2005) A large-scale statistical analysis of barefoot impressions. J Forensic Sci 50(5):1071–1080
Kerstholt JH, Paashuis R, Sjerps M (2007) Shoe print examinations: effects of expectation, complexity and experience. Forensic Sci Int 165:30–34
Koehler JJ (2011) If the shoe fits they might acquit: the value of forensic science testimony. J Empir Legal Stud 8:21–48
Krishan K (2008a) Determination of stature from Foot and its segments in a north Indian population. Am J Forensic Med Pathol 29:297–303
Krishan K (2008b) Establishing correlation of footprints with body weight-Forensic Aspects. Forensic Sci Int 179:63–69
Krishan K (2008c) Estimation of stature from Foot prints and Foot outline dimension in Gujjars of north India. Forensic Sci Int 175:93–101
Krishan K, Kanchan T, Sharma A (2011) Sex determination from hand and foot dimensions in a North Indian population. J Forensic Sci 56:453–459
Laskowski GE, Kyle VL (1988) Barefoot impressions–a preliminary study of identification characteristics and population frequency of their morphological features. J Forensic Sci 33(2):378–388
Lucock LJ (1967) Identifying the wearer of worn footwear. J Forensic Sci Soc 7:62–70
Massey S (2004) Persistence of creases of the foot and their value for forensic identification purposes. J Forensic Identif 54(3):296–315

Mikkonen S, Suominen V, Heinonen P (1996) Use of footwear impressions in crime scene investigations assisted by computerised footwear collection system. Forensic Sci Int 82:67–79
Milne R (2001) Operation Bigfoot–a volume crime database project. Sci Justice 41:215–217
Morgan RM, Freudiger-Bonzon J, Nichols KH et al (2009) The forensic analysis of sediments recovered from footwear. In: Ritz K, Dawson L, Miller D (eds) Criminal and environmental soil forensics. Springer, Dordrecht, pp 253–269
Napier TJ (2002) Scene linking using footwear mark databases. Sci Justice 42:39–43
Naples VL, Miller JS (2004) Making tracks: the forensic analysis of footprints and footwear impressions. Anat Record B New Anatomist 279:9–15
Pataky TC, Mu T, Bosch K et al (2012) Gait recognition: highly unique dynamic plantar pressure patterns among 104 individuals. J R Soc Interface 9:790–800
Pavlou M, Allinson NM (2009) Automated encoding of footwear patterns for fast indexing. Image Vision Comput 27:402–409
Petraco ND, Gambino C, Kubic TA et al (2010) Statistical discrimination of footwear: a method for the comparison of accidentals on shoe outsoles inspired by facial recognition techniques. J Forensic Sci 55:34–41
Pringle JK, Ruffell A, Jervis JR et al (2012) The use of geoscience methods for terrestrial forensic searches. Earth Sci Rev 114(1):108–123
Qamra S, Sharma BR, Kaila P (1980) Naked foot marks – a preliminary study of identification factors. Forensic Sci Int 16:145–152
Robbins LM (1978) The individuality of human footprints. J Forensic Sci 23(4):778–785
Robbins LM (1985) Footprints: collection, analysis, and interpretation. CC Thomas, Springfield
Robbins LM (1986) Estimating height and weight from size of footprints. J Forensic Sci 31:143–152
Rosenbaum D, Becker HP (1997) Plantar pressure distribution measurements. Technical background and clinical applications. Foot Ankle Surg 3:1–14
Rosenbaum D, Hautmann S, Gold M et al (1994) Effects of walking speed on plantar pressure patterns and hindfoot angular motion. Gait Posture 2(3):191–197
Saks MJ, Faigman DL (2008) Failed Forensics: how forensic science lost its way and how it might yet find it. Ann Rev Law Soc Sci 4:149–171. doi:10.1146/annurev.lawsocsci.4.110707.172303
Schallamach A (1968) Abrasion, fatigue, and smearing of rubber. J Appl Polym Sci 12(2):281–293
Sharma BR (1980) Footprints tracks and trails in criminal investigation and trails. Central Law Agency, Allahabad
Sheets HD, Gross S, Langenbug G et al (2013) Shape measurement tools in footwear analysis: a statistical investigation of accidental characteristics over time. Forensic Sci Int 232:84–91
Skerrett J, Neumann C, Mateos-Garcia I (2011) A Bayesian approach for interpreting shoemark evidence in forensic casework: accounting for wear features. Forensic Sci Int 210:26–30
Smith MB (2009) Forensic analysis of footwear impression evidence. Forensic Sci Commun 11(3)
Stone RS (2006) Footwear examinations: mathematical probabilities of theoretical individual characteristics. J Forensic Identif 56:577–599
Thali MJ, Braun M, Buck U et al (2005) VIRTOPSY-scientific documentation, reconstruction and animation in forensics; individual and real 3D data based geo-metric approach including optical body/object surface and radiological CT/MRI scanning. J Forensic Sci 50(2):424–428
Theeuwen ABE, van Barneveld S, Drok JW et al (2001) Enhancement of muddy footwear impressions. Forensic Sci Int 119:57–67
Tuttle RH (1986) Review of 'Footprints. Collection, analysis, and interpretation'. Am Anthropol 88:1000–1002
Tuttle RH (2008) Footprint clues in hominid evolution and forensics: lessons and limitations. Ichnos 15(3–4):158–165
Vernon W (2006) The development and practice of forensic podiatry. J Clin Forensic Med 13(6):284–287
Vernon W, Parry A, Potter M (1999) Moving towards consensus: the first draft of an evaluative instrumental grid to interpret shoe wear patterns. J Forensic Identif 49(2):142–173
Yamashita AB (2007) Forensic barefoot morphology comparison 1. Can J Criminol Crim Justice/La Revue canadienne de criminologie et de justice pénale 49(5):647–656

Chapter 8
Future Directions

Abstract There is something evocative in a human track and for many a trackway provides an instant connection with the track-maker and their journey, blending past with present. Their preservation is in itself a rare occurrence in the geological record and something of a marvel. They contain information not only about human presence, but about the track-makers themselves, as well as the way in which they moved across the landscape providing evidence of their fossilised locomotion. We hope that during the course of this book we have provided information not only on the occurrence of human tracks around the World, but also indicated clearly what can be learned from them, and also what can't. We hope that we have equipped our readers with knowledge of some of the new methods available to enable the study of human tracks whether it is for the sheer pleasure of enquiry, in the pursuit of scientific questions in such fields as geology, geoarchaeology and palaeoanthropology, or in the pursuit of criminals as forensic scientists. In drawing this book to a close we focus on what we consider to be some of the key research themes for the future, questions that we feel need to be addressed by the inter-disciplinary community with an interest in human tracks.

8.1 Future Research Perspectives

Throughout this volume we have attempted to draw together a diverse literature into one place to provide a route map and in some respect a practical manual for those investigating human tracks. Our claim as stated throughout the book is that human tracks are an important part of the geological, archaeological and anthropological record and that these disciplines have a contribution to make in the study of footwear/footprints in a forensic context. We hope that we have demonstrated this and provided a guide for those who have the pleasure and excitement of discovering a new human tracksite or working on an existing one in the future. While human tracksites are a relatively rare occurrence in the geological record, they are there to be found and studied for those with an eagle eye and the knowledge of where to look.

Many of the issues that we have discussed are relevant to the study of other types of vertebrate tracksites whatever the species of track-maker or age, and the influence of substrate on a track is probably similar whatever the shape of the foot that

M.R. Bennett, S.A. Morse, *Human Footprints: Fossilised Locomotion?*,
DOI 10.1007/978-3-319-08572-2_8

made the track. Many of the methodological issues are also common across species of track-maker. It is possible, however, to identify a series of future research perspectives specifically for human tracks, but perhaps also relevant in some cases to the study of vertebrate tracks in general as well, namely:

1. *Methods and technology*. The diversity of methods brought to bear on the study of human tracks is clear from a reading of Chap. 2. As a research community we need to encourage greater standardisation in field methods and data recording across all the diverse component disciplines. This is hard and will not come from any one of the component disciplines but from a convergence of common cause. Robbins (1985) tried to standardise recording methods and we have resisted the temptation to follow this lead by proclaiming our approach to be better than any other; it isn't, but none the less some consensus of approach is needed whether one is dealing with two-dimensional pressure tracks and impressions, three-dimensional tracks or making direct anthropometric measurements. There needs be some form of consistency and minimum standard of data collection allowing greater comparison between sites and studies. Further developments are needed around photogrammetry. Namely to: refine the accuracy and specifically the linear calibration of the three-dimensional models produced in this way; improve the consistency by which models can be made across a range of damp, and therefore often reflective, natural sediment surfaces; and enhance the user interface for the cutting edge freeware which currently gives the best results (Falkingham 2012). Such developments have the potential to revolutionise the collection and study of tracks, placing three-dimensional models into the hand of every geologist, geoarchaeologist, or forensic scientist. However with the increasing availability of good three-dimensional track data we also need to develop new methods of data analysis. Pedobarographic Statistical Parametric Mapping (pSPM) developed by Pataky and his colleagues (Pataky et al. 2008; Crompton et al. 2012) show the potential of such 'whole-foot' analytical tools, but is just one possible approach. Other similar tools are urgently needed and more generally in the authors' view, the number of track studies that apply basic geometric morphometric techniques is limited yet the potential is considerable. We need to develop alternative methods of 'whole-track' analysis and crucially to ensure that any associated software or code is freely available to the community as a whole.
2. *Open Data Access*. One only has to review the large number of papers which explore the empirical relationship of the foot to various body proportions as we did in Sect. 6.2 to realise that despite the inconsistency of method there is a huge amount of data available. The problem is that this data is not freely available. If we were to increasingly pool our data across the different discipline boundaries, creating common data repositories for all researchers to access, then interdisciplinary sciences such as this would advance more rapidly. For example the pooling of anthropometric data across modern racial/ethnic groups would allow us to explore variation within our species as a whole and allow us to develop more generic models. While this may decrease the predictive power of a resultant

model in respect to a particular target population it would give the community as a whole a better foundation for assessing the 'unknown track' and variation within our entire species. There are also intriguing regional variations to be explored with respect to foot anatomy and the tracks produced. Again greater consistency of method and approach to data acquisition would assist here but ultimately it is about the openness with which we chose to share our data within an interdisciplinary community of researchers. Data from fossil tracksites, especially those of greater antiquity, is often closely guarded by the excavators, limiting access for other researchers with different or complementary methods. Competition and rivalry are elements of human nature but they undoubtedly hold back our science. We strongly advocate the open sharing of data and to this end we have put much of the data discussed in this volume on the following site for researchers to access: http://footprints.bournemouth.ac.uk/. We hope others follow our lead making their data available also.

3. *Uniqueness of the human foot and/or track*. How unique is a human track? It is an intriguing question whether you are a forensic scientist or not. Pataky et al. (2012) suggest that plantar pressures may be distinctive to the individual; but to what extent does this translate to the three-dimensional topology of a track? Kennedy et al. (2005) have built a huge repository of data which supports the idea that the probability of two feet matching in terms of their basic dimensions is small; something which they have applied widely in a forensic context. How far does this extend to a three-dimensional track? These are interesting questions for future research. They also lie at the heart of the study of ancient tracks. Tracks with sufficient antiquity to shed light on the locomotion of our ancestors are limited in number; we are often inferring much from the tracks of a single or restricted number of individuals. How representative are they of the population as whole from which they are derived? If they are 'unique' are we not in danger of making false inferences? Understanding the statistical variation in foot morphology and the tracks they leave within a population as a whole become important questions. Bates et al. (2013a) has recently demonstrated that there is a much greater overlap with respect to the flexibility of the mid-foot in some modern humans and the Great Apes than previously supposed. Until we can understand the variability of behaviours within a single population it is hard to predict what might exist in extinct ones. These are very fundamental questions to the study of ancient tracks and not ones that have been adequately explored to date.
4. *Track topology*. Our understanding of the topological variation of human tracks with substrate has improved in the last few years and has applicability across all vertebrate track studies, but there is much more that can be done. The way forward is not necessarily more laboratory experiments, but to make use of real World laboratories on our beaches, estuaries and lake shores because artificial experiments are exactly that, artificial. The interaction of substrate, gait and the resultant tracks is a closely coupled problem with one feeding into the other and we would encourage others to take the lead set by Morse et al. (2013) and Bates et al. (2013b) and explore these interactions further.

These research questions are just some of the potential avenues to explore, we believe that the potential outcomes are worthy of effort and would encourage all those interested in the study of human tracks to seek out new track sites and to develop innovative research agendas around them. The tracks are out there and it is time the authors set aside their key boards, and you this book, and together let's find them!

References

Bates KT, Collins D, Savage R et al (2013a) The evolution of compliance in the human lateral mid-foot. Proc R Soc Lond Ser B280. doi:10.1098/rspb.2013.1818

Bates KT, Savage R, Pataky TC et al (2013b) Does footprint depth correlate with foot motion and pressure? J R Soc Interface 10:1742–5662. doi:10.1098/rsif.2013.0009

Crompton RH, Pataky TC, Savage R et al (2012) Human-like external function of the foot, and fully upright gait, confirmed in the 3.66 million year old Laetoli hominin footprints by topographic statistics, experimental footprint-formation and computer simulation. J R Soc Interface 9:707–719. doi:10.1098/rsif.2011.0258

Falkingham PL (2012) Acquisition of high resolution three-dimensional models using free, open-source, photogrammetric software. Palaeontol Electronica 15:1T:15p

Kennedy RB, Chen S, Pressman IS et al (2005) A large-scale statistical analysis of barefoot impressions. J Forensic Sci 50(5):1071–1080

Morse SA, Bennett MR, Liutkus-Pierce C et al (2013) Holocene footprints in Namibia: the influence of substrate on footprint variability. Am J Phys Anthropol. doi:10.1002/ajpa.22276

Pataky TC, Caravaggi P, Savage R et al (2008) New insights into the plantar pressure correlates of walking speed using pedobarographic statistical parametric mapping (pSPM). J Biomech 41:1987–1994

Pataky TC, Mu T, Bosch K et al (2012) Gait recognition: highly unique dynamic plantar pressure patterns among 104 individuals. J R Soc Interface 9:790–800

Robbins LM (1985) Footprints: collection, analysis, and interpretation. CC Thomas, Springfield

Appendix

Abstract The 12 human tracks that follow were all made by modern humans working at Bournemouth University in 2007 as part of a wider study of 254 subjects. The tracks were collected as described in Sect. 1.4 in a shallow sand tray under normal walking. Tracks were captured digitally via an optical laser scanner (Konica-Minolta, VI900) and the contour maps created from XYZ point clouds within ArcMap. Contour intervals are 1 mm in all cases.

M.R. Bennett, S.A. Morse, *Human Footprints: Fossilised Locomotion?*,
DOI 10.1007/978-3-319-08572-2

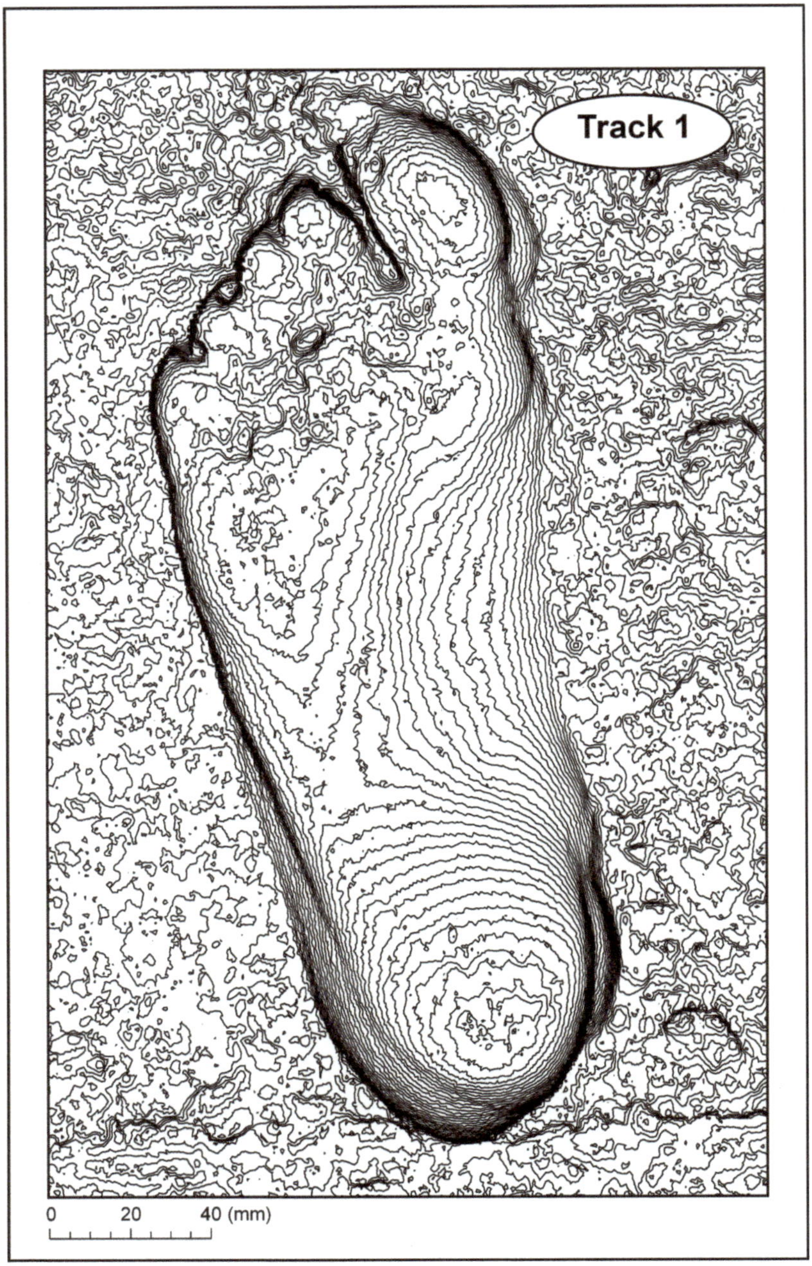

Track-1 [Subject: Female, Height = 1.65 m, Body Mass = 77 kg, Age = 43]. The track is of note and reproduced here because of the width variance around the heel and medially to the first toe. This type of effect helps explain the difference between width measurements made on two-dimensional and three-dimensional tracks and also the variability along a given trackway

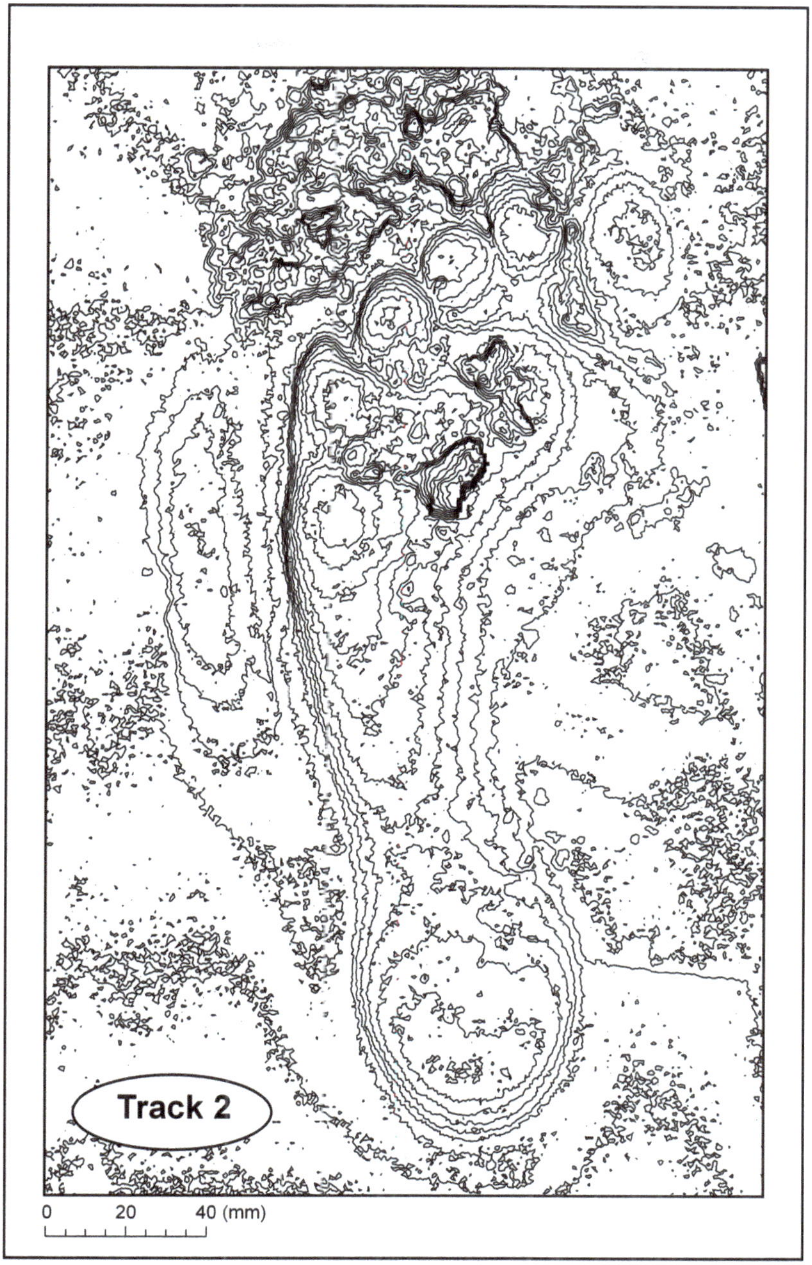

Track-2 [Subject: Female, Height = 1.66 m, Body Mass = 52 kg, Age = 22]. This track shows a supinated foot contact in which the outside lateral edge of the foot bears the weight. This is not a stereotypical pattern of footfall and may be indicative of enhanced mid-foot flexibility in this individual

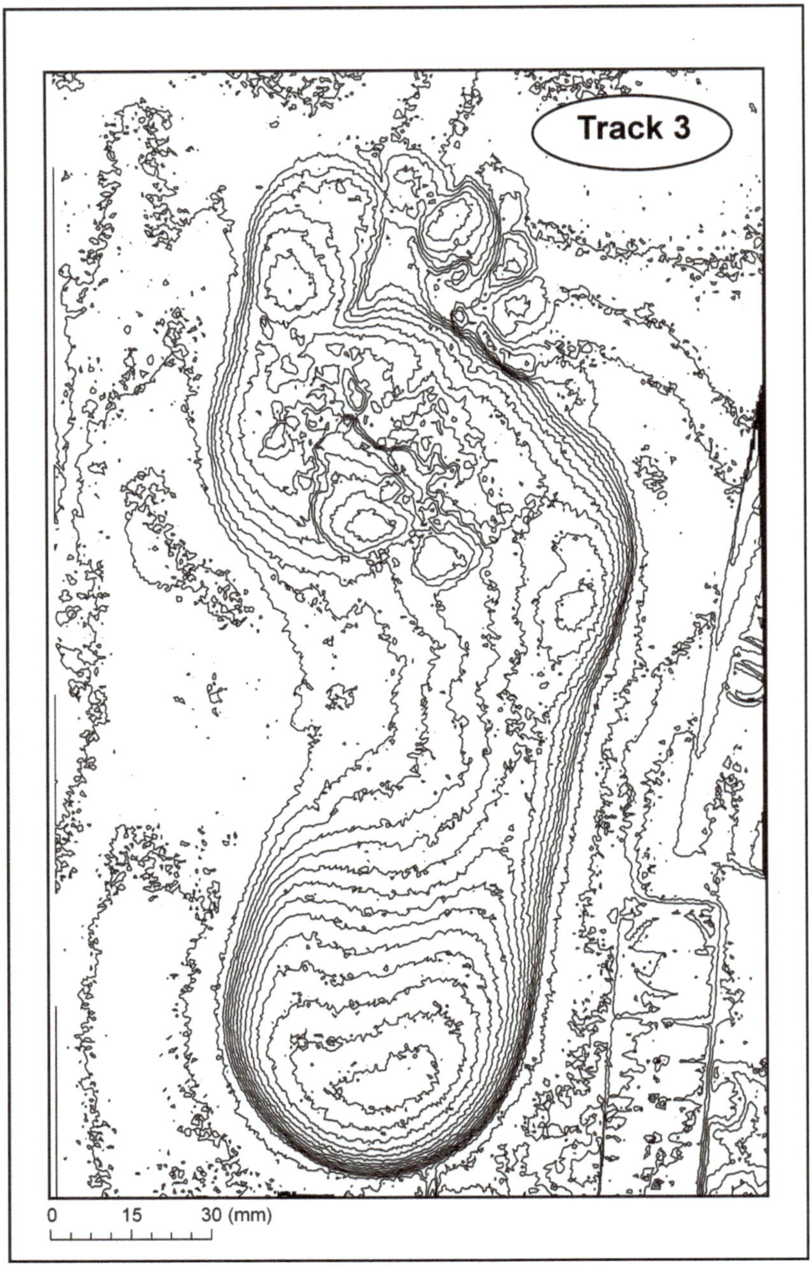

Track-3 [Subject: Female, Height = 1.68 m, Body Mass = 111 kg, Age = 53]. This track is of interest primarily because of the relatively low longitudinal medial arch and the interesting array of toe pads which is typical of the effects of certain types of female footwear. Note the weak second toe pad, elevated above the first and third toes and the bunching of toes pads three to five. Similar toe patterns are not restricted to females but tend to be more common and show the impact of shoes on habitually shod modern feet

Track-4 [Subject: Male, Height = 1.77 m, Body Mass = 75 kg, Age = 44]. This track is reproduced here simply to illustrate the degree to which brittle deformation of thin slabs of displaced sediment may obscure a footprint

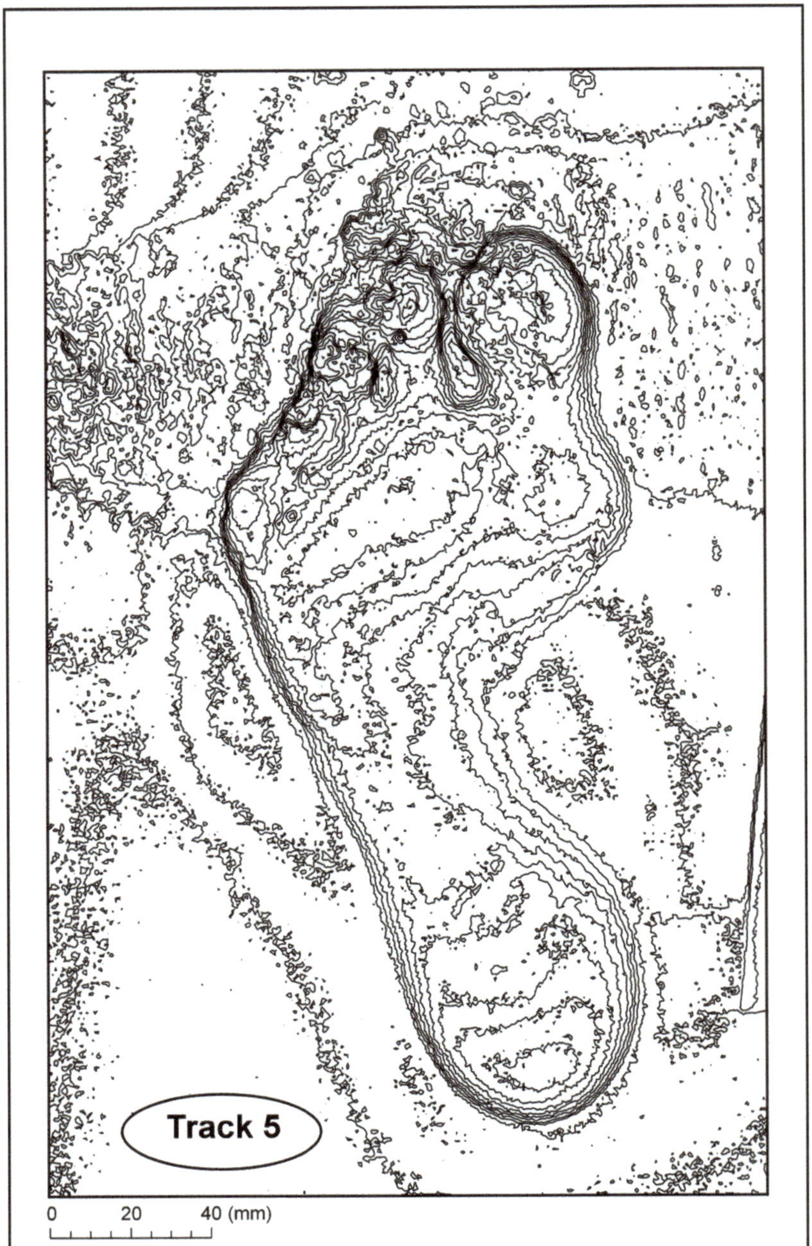

Track-5 [Subject: Female, Height = 1.6 m, Body Mass = 72 kg, Age = 21]. This track shows an extended second toe and also a well-developed and rather prominent bunion. Bunions are not restricted to female feet and may also occur in both males and children. It is however a relatively common feature of female tracks within the Bournemouth data set. In a forensic context a track showing such a feature may be suggestive although not a definitive indicator of sex

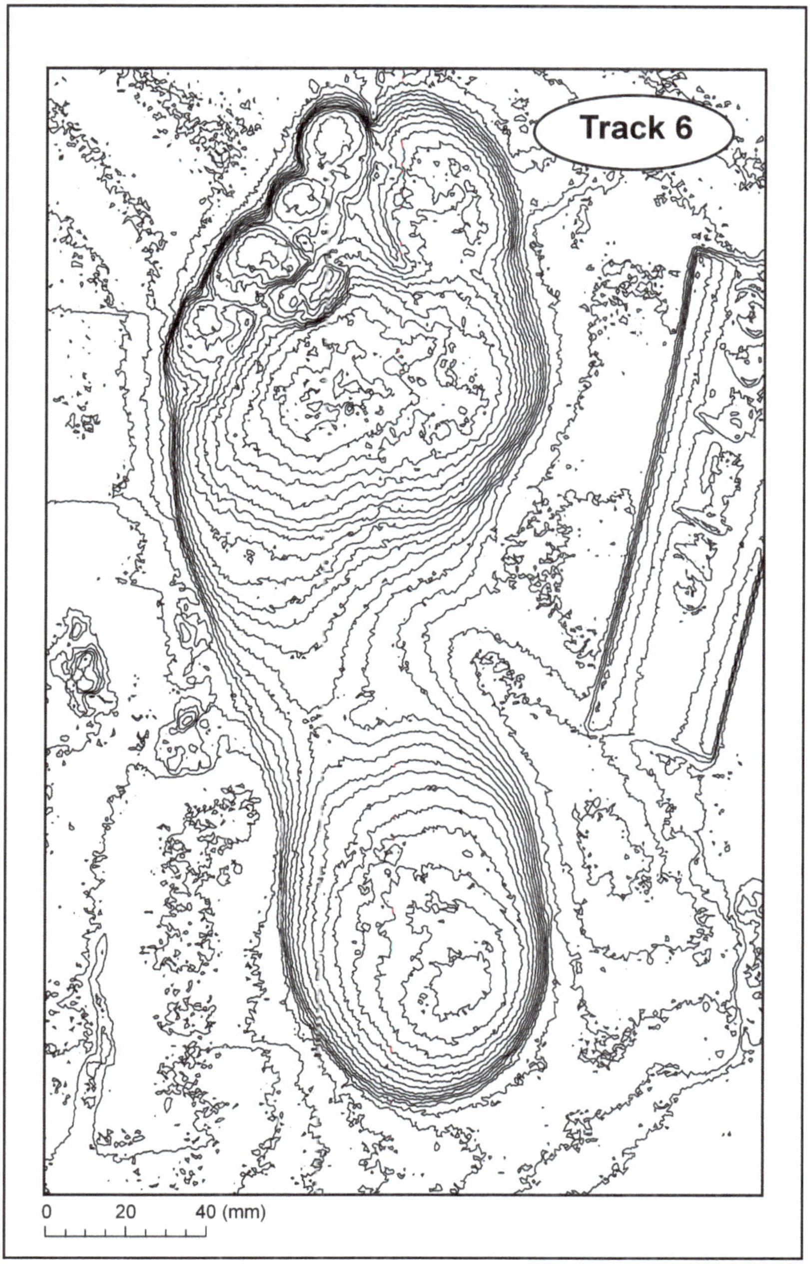

Track-6 [Subject: Female, Height = 1.54 m, Body Mass = 54 kg, Age = 32]. Typical track showing an extended second toe and a well-developed longitudinal medial arch

Track-7 [Subject: Male, Height = 1.72 m, Body Mass = 82 kg, Age = 43]. The track is of note primarily because of the two phases of tectonics in the region of the toes/ball. Initial movement of the lesser toes in a medial direction transverse to the long axis of the track has distorted the lateral outline of the track and generated a ridge of sediment running down the long axis of the foot. This is cross-cut during a second phase by movement directed anteriorly from behind the first toe during the final phase of toe-off

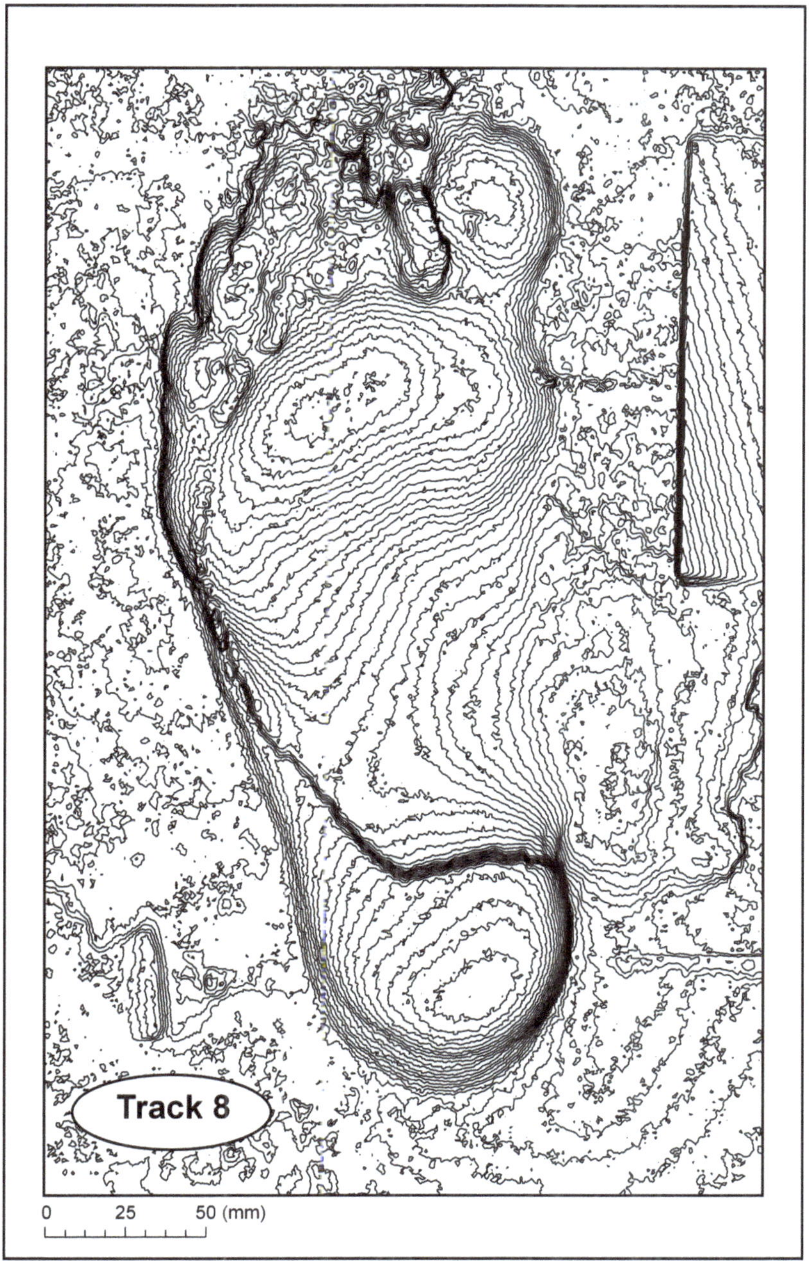

Track-8 [Subject: Male, Height = 1.66 m, Body Mass = 54 kg, Age = 21]. This track shows a typical deformation structure to the rear of the mid-foot. Note that the first and the second metatarsal heads can be seen in the track

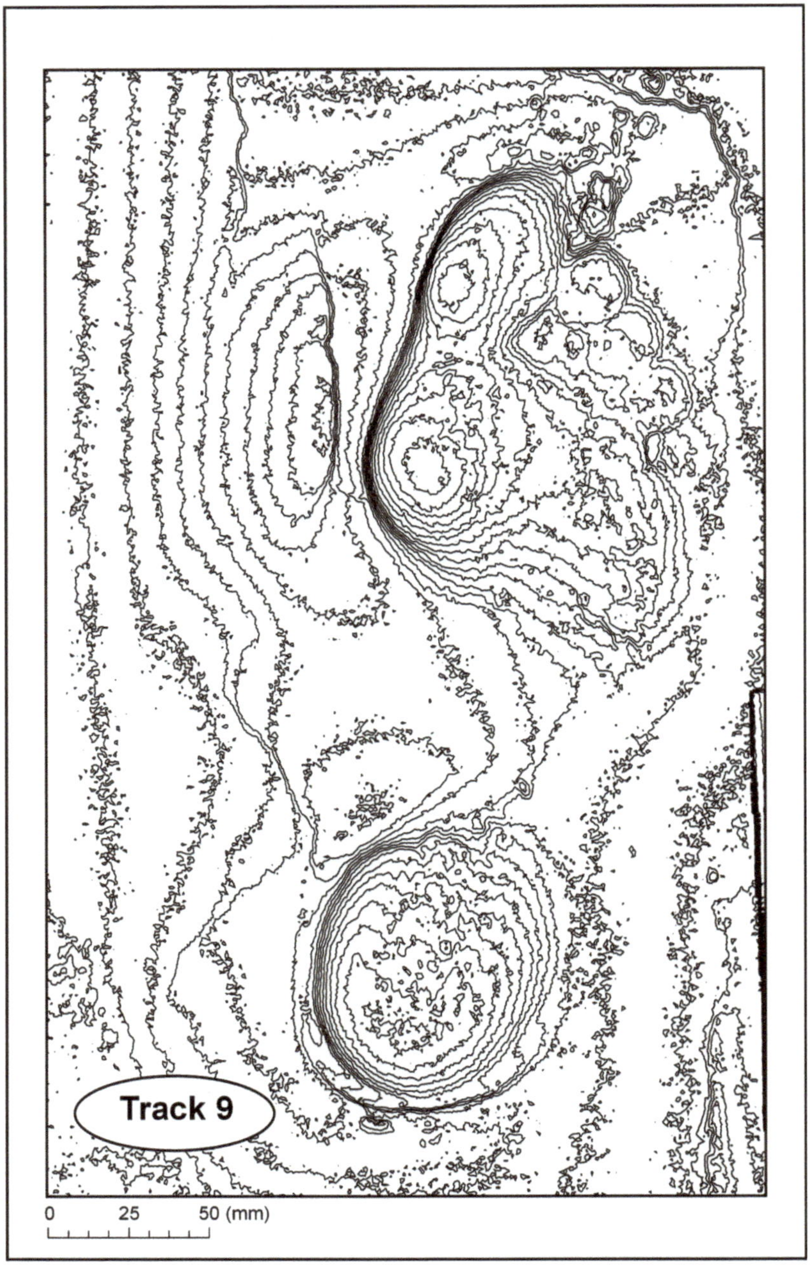

Track-9 [Subject: Female, Height = 1.6 m, Body Mass = 72 kg, Age = 21]. This track shows a very well-developed longitudinal medial arch, which causes complete separation of the heel from the ball areas of the foot. Note also the prominent medial transfer of weight leading to the deepest point of the ball being on extreme medial side of the foot. Pressure on the medial track wall in this region has caused the formation of a small rim structure

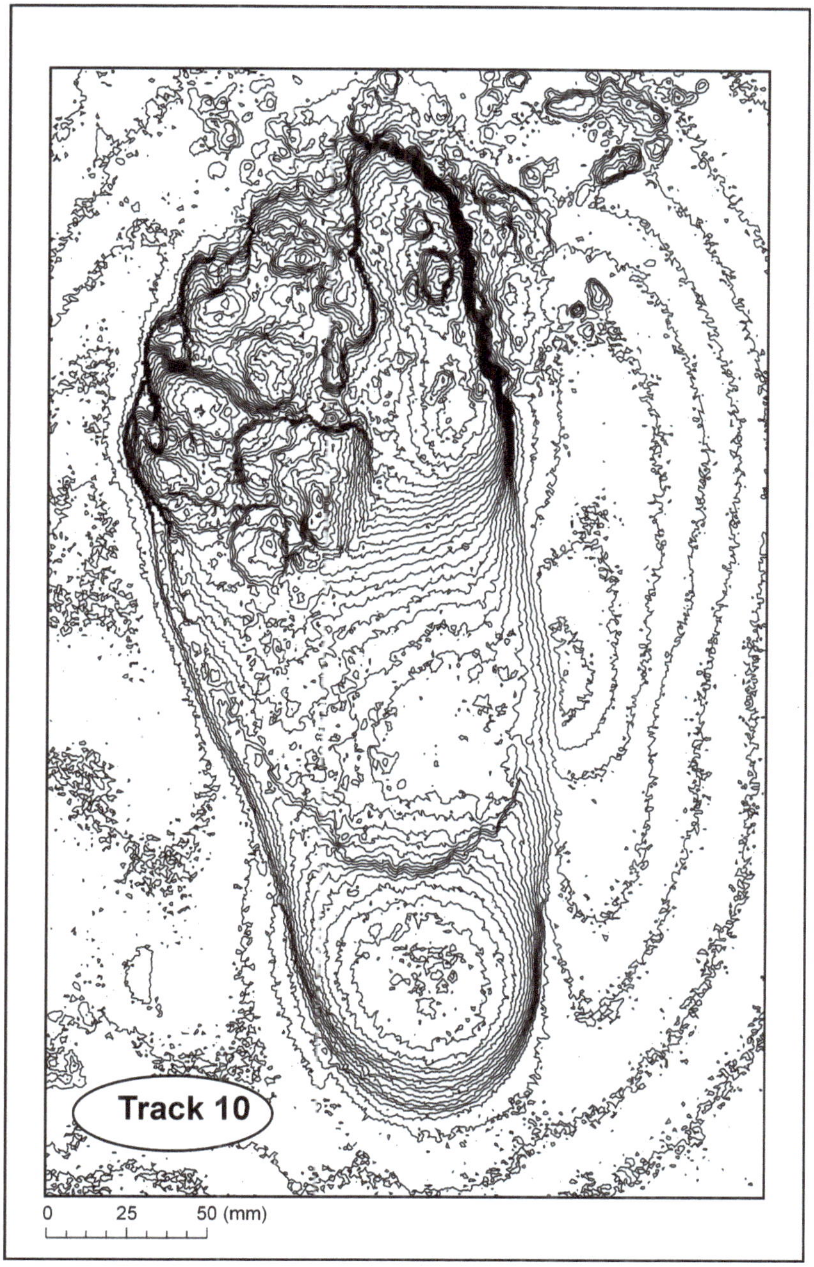

Track-10 [Subject: Female, Height = 1.6 m, Body Mass = 165 kg, Age = 26]. The track shows how the longitudinal medial arch may be completely obscured by the anterior movement of sediment from behind the ball of the foot. Note the prominent scarp formed by this sediment within the heel area. Such structures have been used in fossil prints to attest to the presence of enhanced mid-foot flexibility, which is not true in this case

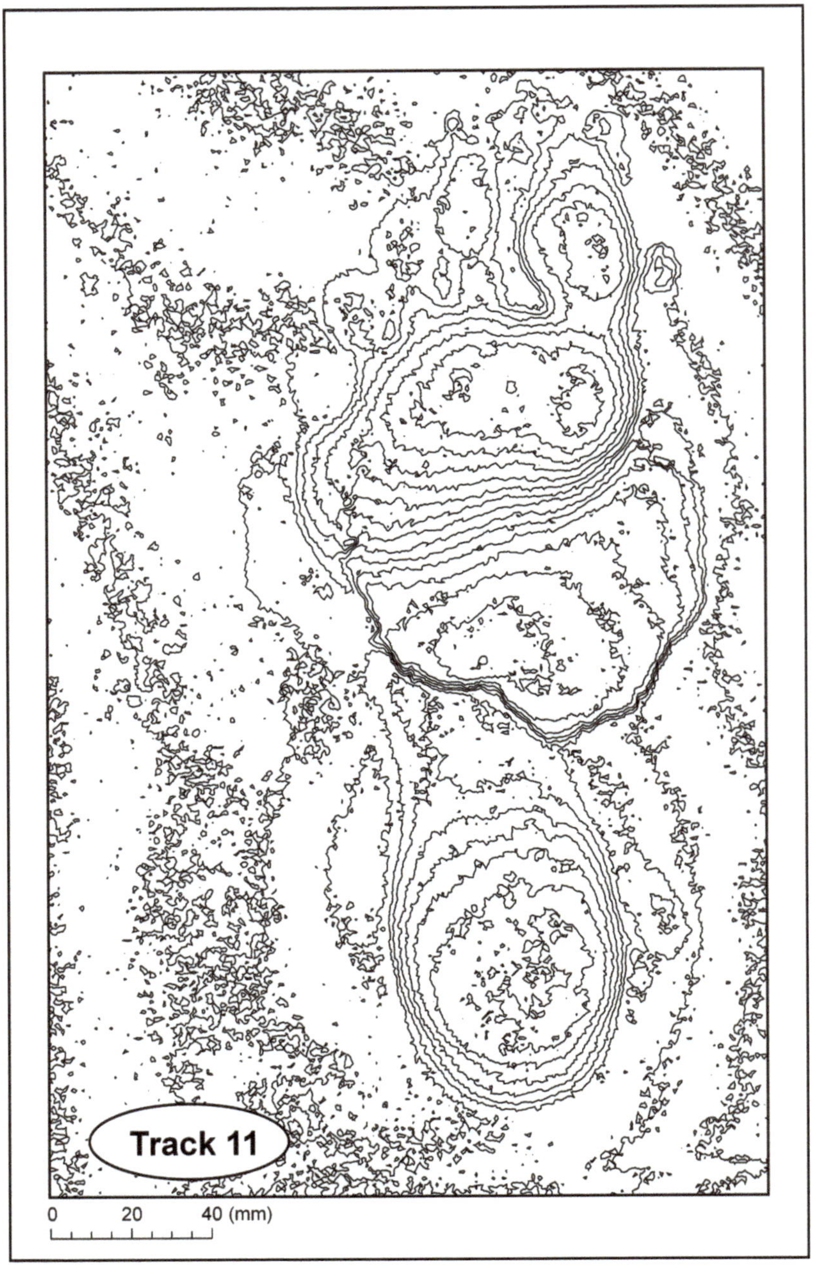

Track-11 [Subject: Male, Height = 1.85 m, Body Mass = 94 kg, Age = 36]. Note the deepest point in the ball is below the first and second metatarsal heads which resulted in the displacement of a shallow sediment slab in an posterior direction impacting on the medial boundary of the track. Note the poor definition of the less toe pads

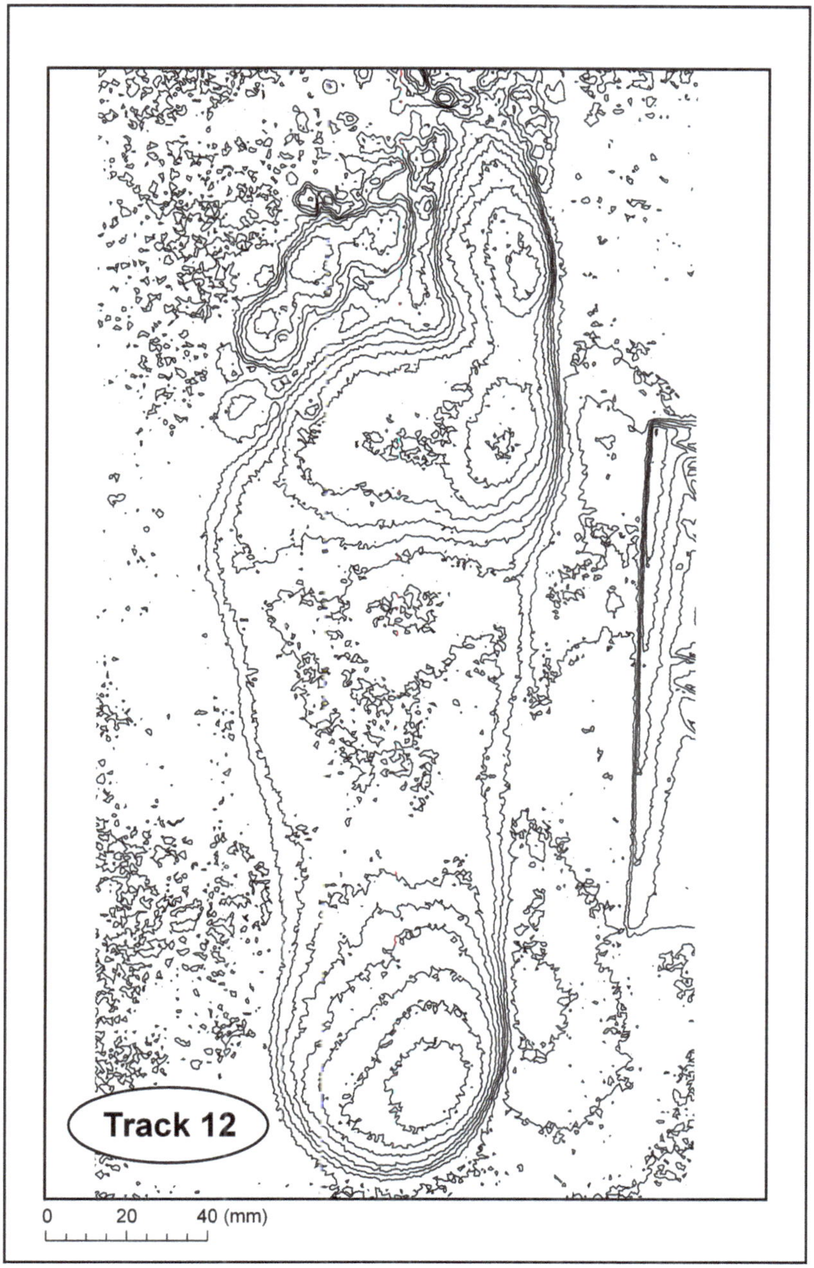

Track-12 [Subject: Male, Height=1.5 m, Body Mass=42 kg, Age=12]. This track shows a poorly developed medial longitudinal arch, prominent medial toe-off the combined effect is to create a very linear medial edge to the track

Glossary

Abduction The movement of the first digit of the foot away from the longitudinal axis of the foot and the opposite of adduction. [Sect. 1.2]

Adduction The alignment of the first digit of the foot with the longitudinal axis of the foot and the opposite of abduction. [Sect. 1.2]

Ages and dating conventions In this volume we use the nomenclature Ma to refer to millions of years and ka for thousands of years. The initials BP refer to Before Present which is defined as 1 January 1950 and is a convention used when dealing with radiocarbon dates. [Sect. 2.3]

Allometric (allometry) The relative size and shape of different component parts of the body and their variation during an organism growth. [Sect. 6.2]

Anthropometric (anthropometry) Measurements of the human body as an aid to anthropological classification and comparison. [Sect. 6.2]

Artiodactyla An order of hooved mammals which have an even number of functional toes.

Bent hip, bent knee A style of walking that involves locomotion with both a flexed knee and hip and is common to some primates and has also been assigned by some authorities to early hominin species. [Sect. 6.4]

Bilateral asymmetry Differences in size and/or shape between the right and left sides of the body that is either side of the Sagittal Plane. [Sect. 6.2]

Bipedalism Upright walking. [Sect. 1.2]

Body Mass Index (BMI) A method of combining an individual's height and weight to define broad characteristics such as underweight, overweight or obese. It corresponds to the mass divided by the square of height. Classifications vary regionally but in general a BMI of over 30 is considered to be obese while one of 15 corresponds to someone who is severely underweight. [Sect. 6.2]

Bourrelet A terms used in some vertebrate track studies to refer to a rim-structure or a push-up structure around the margin of a track. [Sect. 1.2, Table 1.1]

Bulk density The mass of sediment or soil in a known volume. Refers generally to the particle packing, porosity and permeability of sediment. [Sect. 5.2]

The purpose of this glossary is to help those who are not familiar with the jargon of one or more the component disciplines. It is not intended to be exhaustive, simply a guide to assist the reader navigate some of the terms used in this book.

M.R. Bennett, S.A. Morse, *Human Footprints: Fossilised Locomotion?*,
DOI 10.1007/978-3-319-08572-2

Congregation site A type of track site in which tracks appear to concentrate around a focal point such as a watering hole in contrast to a transit site where they pass across a site. [Sect. 1.2]

Digital elevation models A digital model of surface terrain usually defined by a series of x, y and z coordinates. In the context of this book it refers to the micro-terrain of a track. [Sect. 2.4]

Dorsal Upper surface of the foot, opposite of plantar. [Sect. 1.2]

Dorsiflexion Movement of the foot that raises the foot or toes upwards extending the plantar surface and contracting the dorsal surface of the foot; opposite of plantarflexion. [Sect. 1.2]

Eversion A turning outward of the foot; opposite of inversion. [Sect. 1.2]

Foot index A ratio of foot length to breadth. [Sects. 2.6 and 6.2]

Froude number A dimensionless number defined as the ratio of a characteristic velocity to a gravitational wave velocity. It is used widely within fluid mechanics but has also been used to characterise human and animal walking. In this case the leg is considered to be equivalent to a pendulum. [Sect. 6.4]

Gait A manner of moving the foot in order to walk or run. [Sects. 2.6 and 6.4]

Generalised Procrustes analysis In Greek mythology Procrustes stretched people or cut off their limbs in order to fit them to a standard sized iron bed. It refers now to a mathematical approach to the description of shape in which landmarks are transformed via scaling, translation and rotation to fit a common form. In theory it provides a method of studying shape free from the influence of size. [Sect. 2.7, Fig. 2.16]

Geoheritage (geoconservation) The conservation of geological phenomena such as rocks, fossils, tracks and geomorphological processes. Geotourism is a general term used to describe tourism activity that has a geological focus, for example visitors coming to see a series of tracks or an aesthetically pleasing landscape. [Sect. 4.1]

Geometric morphometrics The statistical representation and analysis of anatomical shape. [Sect. 2.7]

Greater trochanter The trochanter is one of several bony protuberances at the top end of the thigh bone or femur. The greater trochanter is located on the proximal and lateral part of the shaft of the femur and is also sometimes referred to as the major trochanter or outer trochanter. For most purposes within gait studies it is taken as the height of the hip. [Sect. 6.4]

Hallux The bones of the first digit of the foot sometimes referred to as hallucal. [Sect. 1.2]

Hominins Refers to a creature that is either human or an agreed human ancestor, note that this term has replaced the previous one of hominid reflecting a broader definition of ancestor. This typically includes all of the *Homo* species (*Homo sapiens*, *H. ergaster*, *H. rudolfensis*), all of the *Australopithecines* (e.g., *Australopithecus afarensis*) and other ancient forms like *Paranthropus* and *Ardipithecus*.

Hydrogeology The study of groundwater and the geology associated with it. In the current context the term refers to the geometry and height for example of the water table and its influence of sediment moisture content. [Sect. 5.3]

Ichnofossil A trace or mark left by an animal that has been preserved (fossilised) within the geological record. [Sect. 1.2]

Ichnology The study of trace fossils. [Sect. 1.2]

Ichnotaxonomy The taxonomy of trace fossils. [Sect. 1.2]

Ipsilateral A line joining the Pternion (anterior point of the heel) on one track with the Pternion of same track made by the next footfall. That is a line joining the heel points of successive left or right feet in a trackway. [Sect. 2.6, Fig 2.11]

Lagomorphs Members of the taxonomic order Lagomorpha, of which there are two living families: the Leporidae (hares and rabbits) and the Ochotonidae (pikas). Rabbits and hares characteristically have long ears, a short tail, and strong hind limbs that provide a bounding locomotion. In contrast, the smaller pikas have shorter, rounded ears, no external tail, and less-well-developed hind limbs associated with scampering locomotion. [Sect. 3.1.1]

Landmarks A landmark is a point placed on a specimen in this case a track for the purpose of taking measurements or defining a point of interest. Landmarks may be two points between which a linear measurement is taken, or they may be defined specifically by either two-dimensional or three-dimensional coordinates. Homology-based landmarks relate to biologically or anatomically structures that can be recognised consistently by observers, for example the tip of the first toe or the most anterior point of the heel. Homology-free landmarks are determined by some form of geometrical or mathematical principle. Sliding-landmarks or semi-landmarks are landmarks that are placed on smooth surfaces and allowed to move in order to optimise the description of a curve or surface. [Sects. 2.6 and 2.7]

Lateral The outside edge of the foot furthest away from the medial plane of the body and the opposite of medial. [Sect. 2.1]

Liquefaction Process by which a sediment or solid become liquefied usually through the addition of water and through vibration. [Sect. 5.2]

Lithification The process by which sediment is turned to stone or lithified. This process may involve the transformation of the sediment in which case the term diagenesis is used. [Sect. 5.4]

Medial The inside edge of the foot closest to the medial plane of the body and the opposite of lateral. [Sect. 1.2]

Metatarsal heads The expanded distal end of a metatarsal bone as it joins to the proximal phalanx of the same digit. In lay terms it corresponds approximately to the ball of the foot. [Sect. 1.2, Fig. 1.2]

Microbial (algal) mat A layer of algal filaments which form a surface covering over damp sediment, often covering a track. They aid sedimentation by providing a baffle interrupting sediment transport and may also contain sticky surfaces to which sediment may adhere. [Sect. 5.3 and 5.4, Fig. 5.15]

Mid-foot The area of the foot between the ball and the heel, corresponding broadly to the instep. [Sect. 1.2, Fig. 1.2]

Mid-tarsal break A characteristic of certain apes and some humans in which the mid-foot shows greater mobility. Essentially instead of forming a rigid leaver from the ball of the foot, the foot flexes in the mid-foot region especially on the lateral side. [Sect. 6.4]

Neoichnology The study of trace fossils left on modern surfaces by known track-makers, effectively providing modern analogue information. [Sects. 1.2 and 5.1]

Optical laser scanner A devise which uses a laser to provide a detailed three-dimensional image of a surface usually by producing a series of x, y and z coordinates for a closely spaced area. Scanners work on different principles but most involve calculating the return time and associated angles of a laser beam. [Sect. 2.4]

Overall track A term used to describe the overall size of a track when viewed from above. If a track has flared walls this may be larger than actual area covered by the track bottom or base and therefore the foot that made the track. [Sect. 1.2, Table 1.1]

Overprinting A track superimposed over another partially or completely destroying the first track. [Sect. 1.2, Table 1.1]

Palaeosurface An ancient landsurface usually in this context containing tracks. [Sect. 1.2]

Photogrammetry The process of triangulation between two or more photographs and a fixed point on those images that allows the relative position (x, y and z) of that point to be defined. It provides a method by which digital elevation models of tracks can be calculated by matching pixels on one or more photographs. [Sect. 2.4]

Plantar The sole of the foot or pertaining to the sole of the foot. [Sect. 1.2]

Plantarflexion Movement of the foot that flexes the foot or toes downward extending the dorsal surface of the foot and contracting the plantar surface; opposite of dorsiflexion. [Sect. 1.2]

Plantigrade Walking with the foot flat on the ground. In contrast to digitigrade where walking occurs on the toes with the heel raised and unguligrade which is walking on the nail or nails (hoof) of the toes with the heel and the digits permanently raised. [Sect. 1.2]

Point clouds A general term used to describe point based three-dimensional data. A three-dimensional object can be represented by lots of points each with an individual x and y coordinate, as well as a z value that records the height of the point with respect to a reference datum. This reference datum may be the position of a scanner, camera or some other reference datum. If the points are plotted they form a cloud which represents the three-dimensional form of that surface. These points can joined and/or represented in various ways to give a three-dimensional representation of a surface and can be used to calculate such things as contour lines (lines of equal height). [Sect. 2.4]

Pore-water content Moisture held in the pores or spaces within sediment also more loosely referred to as sediment moisture content. [Sect. 5.2]

Proboscidea An order of massive herbivorous mammals that have tusks and a long trunk such as elephants.

Pronation Rotation of the foot (or hand) so that the inside or medial edge bears the body's weight; opposite of supination. [Sect. 1.2]

Pternion Most anterior point on the heel. [Sect. 2.6, Fig. 2.9]

Pyroclastic flow A hot (usually) and violent outburst or flow of volcanic ash during a volcanic eruption. It is typical of a Plinian eruption which is particularly violent. [Sects. 3.1.5 and 3.2.7]

Rankine shear zones A particular theoretical depiction of the likely failure surfaces (slip planes) induced in a sediment by vertical loading, in this case by a foot. [Sect. 5.2, Fig. 5.5]

Regions on interest Refers to areas of anatomical interest in this case to areas of a track or the foot. It is intermediate between a landmark based study and a 'whole-track' study. [Sect. 2.7]

Rheology The ductile properties in this case of sediment. [Sect. 5.2]

Sagittal Plane Sagittal Plane is a vertical plane which passes from front to the rear of the body dividing it into right and left halves. It is perpendicular to the Coronal Plane which divides the body front from rear. [Sect. 1.2]

Sediment consistency A general descriptive term for a range of sediment properties (e.g., grain size, sorting, permeability, porosity and pore-water content) which describe in general terms the rheology of a sediment. [Sect. 5.2]

Sediment sorting The range of grain sizes within an admixture of particles. A poorly sorted sediment is one with a wide range of different sizes present whereas a well sorted one has just a few. [Sect. 5.2]

Sexual dimorphism The size differences and/or relationship between the sexes of a species. In this context the difference in size and shape between tracks made by males and females. [Sect. 6.2]

Shear stress The external force acting on an object or surface parallel to the slope or plane on which it rests. In more general terms stress is the force applied over an area of an object and may induce it to shear or fail. [Sect. 5.2]

Soft-sediment deformation Structure that develops within and between sedimentary units during or immediately after deposition. The beds begin to inter-mix to varying degrees due to differential loading, for example of heavier units above finer more saturated ones, or due to ground disturbances caused by such things as earthquakes of slumping of sediment on depositional slopes. Various different structures from simple flame structures to ball and pillow forms to highly convoluted bedding. [Sect. 2.2]

Strain The deformation, in this case of sediment, as a result of an applied stress and refers to its total deformation (compression, expansion, extension) in response to the applied stress. [Sect. 5.2]

Substrate Generic descriptor for any organic material, sediment, rock or combination thereof over which a track-maker walks. [Sect. 5.2]

Supination Rotation of the foot (or hand) so that the outer or lateral edge bears the body's weight. Opposite of pronation. [Sect. 1.2]

Survivorship Refers to the probability that a track will survive without being destroyed and stand a change of being fossilised. A high survivorship refers to a track or conditions that are likely to lead to the burial and/or fossilisation of a track. [Sect. 5.4]

Taphonomy The study of the conditions and processes by which something become fossilised, in this case the formation, preservation and fossilisation of a human track. [Sect. 5.4]

Thixotropy A property exhibited by some fine-grained sediments, mainly clays, in which they become thinner and less viscous when disturbed or shaken. [Sect. 5.2]

Track ejecta Debris ejected from a track with the withdrawal of a foot. [Sect. 1.2, Table 1.1]

Track topology The three-dimensional surface relief of the deformed substrate which forms the track. Track morphology – form and structure of an item – is also sometime used. Both pertain to the physical shape, geometry and size of a track. [Sect. 1.2, Table 1.1]

Track A single footprint or partial impression made by the foot of an animal. [Sect. 1.2, Table 1.1]

Track-maker The animal responsible for making a track or trackway. [Sect. 1.2, Table 1.1]

Trackway A series of tracks made by the same animal. [Sect. 1.2, Table 1.1]

Transit site A type of track site in which the tracks record the movement of humans and/or animals across a site in contrast to a congregation site in which the cluster. [Sect. 1.2]

Tufa A deposit usually of calcium carbonate deposited from freshwater (or saline water) layers of tufa may build up to give travertine which is a freshwater limestone. It is common in areas rich in limestone and chalk and those with carbonate rich groundwater especially those emerging under pressure. [Sect. 3.2.5]

Tuff (tephra) A layer of lithified volcanic ash, also referred to as tephra. This ash layer may be deposited in a variety of different ways during a volcanic eruption such as via airfall or alternatively it may be re-worked by slope and/or fluvial processes. It generally represents in geological terms an isochronous surface, that is one of equal age, sometimes referred to as an event horizon. [Sect. 2.3]

Ungulate A diverse group of large mammals composed of the hoofed mammals such as horses, cattle, deer, swine, and elephants. The Ungulate order is now divided into the orders Perissodactyla and Artiodactyla.

Whole-track analysis This is a type of analysis in which the whole topology of a footprint is used rather than summarising it via linear measurements of landmarks. It involves co-registering a series of tracks such that anatomically similar points are registered one above another; in crude terms one track is transformed so that it lies in the same coordinate space and is the same size as another track. The only method currently available for this type of analysis is Pedobarographic Statistical Parametric Mapping (pSPM) which provides both an automated and a manual registration process by which tracks can be co-registered thereby allowing measures of central tendency such as mean tracks to be calculated. It is also possible to compare two tracks statistically pixel by pixel; each pixel representing in the case of a footprint depth. [Sect. 2.7]

Geographical Index

M.R. Bennett, S.A. Morse, *Human Footprints: Fossilised Locomotion?*,
DOI 10.1007/978-3-319-08572-2

Kolibri 360 GmbH
Ferdinand-Jühlke-Straße 7
99095 Erfurt, Deutschland
produktsicherheit@kolibri360.de